3-D Seismic Interpretation

3-D seismic data have become the key tool used in the oil and gas industry to understand the subsurface. In addition to providing excellent structural images, the dense sampling of a 3-D survey can sometimes make it possible to map reservoir quality and the distribution of oil and gas. The aim of this book is to help geophysicists and geologists new to the technique to interpret 3-D data while avoiding common pitfalls.

Topics covered include basic structural interpretation and map-making; the use of 3-D visualisation methods; interpretation of seismic amplitudes, including their relation to rock and fluid properties; and the generation and use of AVO and acoustic impedance datasets. Also included is the increasingly important field of time-lapse seismic mapping, which allows the interpreter to trace the movement of fluids within the reservoir during production. The discussion of the acquisition and processing of 3-D seismic data is intended to promote an understanding of important data quality issues. Extensive mathematics has been avoided, but enough detail is included on the effects of changing rock and fluid properties to allow readers to make their own calculations.

The authors of *3-D Seismic Interpretation* are professional geophysicists with many years' experience in the oil industry. They are still actively interpreting 3-D seismic data and are therefore able to summarise the current best practice. The book will be indispensable for geoscientists learning to use 3-D seismic data, particularly graduate students of geophysics and petroleum geology, and new entrants into the oil and gas industry.

Mike Bacon was awarded a Ph.D. in geophysics from the University of Cambridge before becoming a Principal Scientific Officer at the Institute of Geological Sciences in Edinburgh (now the British Geological Survey). After working as a lecturer in the Geology Department of the University of Accra, Ghana, he took a position with Shell UK where he worked for 19 years as a seismic interpreter and as team leader in seismic special studies. Dr Bacon is a co-author of *Introduction to Seismic Interpretation* by McQuillin *et al.* (1979) and is a member of the editorial board of the petroleum industry magazine *First Break*. He is a Fellow of the Geological Society and a member of the EAGE (European Association of Geoscientists and Engineers).

Rob Simm is a geophysicist with 16 years' experience in the oil and gas industry and a specialist in the rock physics interpretation of seismic data in both exploration and production. After gaining an M.Sc. and Ph.D. in marine geology at University College London, the early part of his career was spent with Britoil plc and Tricentrol plc as a seismic interpreter. He subsequently took a position at Enterprise Oil and progressed from North Sea exploration to production and equity determination, prior to becoming an internal consultant to asset teams and management. Since 1999 Dr Simm has provided independent consultancy and training services to numerous independent and multi-national oil companies through his company Rock Physics Associates Ltd.

Terry Redshaw gained a Ph.D. in numerical analysis from the University of Wales before becoming a Geophysical Researcher with Western Geophysical. Since 1985 he has been employed by BP in a variety of roles. These have included research into imaging and inversion algorithms, as well as leading a team supplying BP's worldwide assets with support in the areas of seismic modelling, rock properties, AVO and seismic inversion. Dr Redshaw works at present in BP's Exploration Excellence team, which helps operating units to carry out the technical work needed to evaluate oil prospects and decide whether to drill them or not.

3-D Seismic Interpretation

M. Bacon

R. Simm

T. Redshaw

PUBLISHED BY THE PRESS SYNDICATE OF THE UNIVERSITY OF CAMBRIDGE
The Pitt Building, Trumpington Street, Cambridge, United Kingdom

CAMBRIDGE UNIVERSITY PRESS
The Edinburgh Building, Cambridge CB2 2RU, UK
40 West 20th Street, New York, NY 10011-4211, USA
477 Williamstown Road, Port Melbourne, VIC 3207, Australia
Ruiz de Alarcón 13, 28014 Madrid, Spain
Dock House, The Waterfront, Cape Town 8001, South Africa

http://www.cambridge.org

First published 2003

Printed in the United Kingdom at the University Press, Cambridge

Typefaces Times 10.5/14 pt. and Helvetica Neue   *System* LaTeX 2_ε [TB]

A catalogue record for this book is available from the British Library

Library of Congress Cataloguing in Publication data

Bacon, M. (Michael), 1946–
3-D seismic interpretation / by M. Bacon, R. Simm, T. Redshaw.
       p.   cm.
Includes bibliographical references and index.
ISBN 0 521 79203 7 (hardback)
1. Seismic reflection method.   2. Seismic prospecting.   3. Petroleum – Geology.
4. Natural gas – Geology.   I. Title: Three-D seismic interpretation.   II. Simm, R. (Robert), 1959–
III. Redshaw, T. (Terence), 1957–   IV. Title.
QE539.B24   2003
622′.1592–dc21   2003041201

ISBN 0 521 79203 7 hardback

Contents

6 Inversion

7 3-D seismic data visualisation

8 Time-lapse seismic

Preface

Applied geophysics uses a large number of methods to investigate the subsurface. Because of its ability to produce images down to depths of thousands of metres with a resolution of tens of metres, the seismic method has become by far the most commonly used geophysical method in the oil and gas industry. In the past 20 years, the quality of seismic information has been greatly improved by the use of 3-D seismic methods. However, extracting useful information from seismic images remains the interpreter's craft skill, in which elements of geological and geophysical knowledge are combined in varying proportions. This book is intended for people beginning to develop that skill, either as part of a University course or at the beginning of a career in the oil and gas industry. It assumes that the reader has some general background knowledge of the seismic method. There are several excellent texts that cover the whole range of theory and practice (for example, R. E. Sheriff & L. P. Geldart, *Exploration Seismology* (2nd edn, 1995), Cambridge University Press). Our intention is not to replace these volumes, but rather to concentrate on the techniques of interpretation that are specific to 3-D seismic, or are greatly improved in usefulness by applying them to 3-D datasets (such as amplitude studies, AVO analysis, inversion and time-lapse seismic). However, there is enough explanation of the underlying principles to make the book fairly self-contained. In particular, the acquisition and processing of 3-D seismic data are described in some detail. This is partly because the interpreter needs to understand the limitations of his or her data, and whether misleading artefacts are likely to exist in the images that reach his or her desk. Also, he or she will sometimes need to interact with specialists in acquisition and processing, so should understand something of their specialised language. Bearing in mind the diversity of academic background among potential readers, we have avoided any extensive use of mathematics.

The range of topics that might be included is large, and we have tried to concentrate on those that are of most practical application in the authors' experience. There have been rapid advances in interpretation techniques over the past decade. In part this reflects the availability of more computer power at the desktop, so that first-pass interpretations can now often be made in days rather than months. At the same time, data quality has been improving, so that a wealth of detailed subsurface information can be extracted if the right methods are used. We have tried to portray the current state of the art in both these respects. The combination of the interpreter's ingenuity with even more computer power will surely lead to further developments in the future.

We have included a number of examples of seismic displays to illustrate the various interpretation techniques, and to give the reader a feeling for the typical quality of modern seismic data. We are grateful to the following for permission to reproduce proprietary or copyright material: BP Exploration for figs. 2.2, 2.8, 2.16, 2.23–2.24, 2.27, 2.30, 2.34–2.37, 8.3 and 8.7–8.8; ChevronTexaco and Statoil for fig. 5.12; Shell UK Exploration and Production for figs. 3.1, 3.3, 3.5–3.6, 3.8–3.13, 3.17–3.18, 3.20–3.24, 4.4, 4.6, 5.6, 6.2–6.8 and 6.10; the Wytch Farm partnership (BP Exploration Operating Co Ltd, Premier Oil plc, Kerr McGee Resources (UK) Ltd, ONEPM Ltd and Talisman North Sea Ltd) for figs. 7.1–7.6; the Geological Society of London and Dr R. Demyttenaere for fig. 1.6(b); the McGraw-Hill Companies for fig. 5.3; the European Association of Geoscientists and Engineers (EAGE) and Dr J. Hendrickson for fig. 5.16; the EAGE and Dr P. Hatchell for figs. 8.4–8.5; the EAGE and Dr J. Stammeijer for fig. 8.6; the Society of Exploration Geophysicists (SEG) for fig. 4.1, the SEG and Dr S. M. Greenlee for fig. 1.6(a), the SEG and Professor G. H. F. Gardner for fig. 5.1, the SEG and Dr H. Zeng for fig. 4.7, the SEG and Dr W. Wescott for fig. 4.8, and the SEG and Dr L. J. Wood for fig. 4.9. Figures 3.1, 3.3 and 3.24 were created using Landmark Graphics software, fig. 4.6 using Stratimagic software (Paradigm Geophysical), fig. 5.15(b) using Hampson–Russell software and fig. 6.3 using Jason Geosystems software.

The text is intended as an aid in developing understanding of the techniques of 3-D interpretation. We have not been able to include all the possible limitations on applicability and accuracy of the methods described. Care is needed in applying them in the real world. If in doubt, the advice of an experienced geophysicist or geologist should always be sought.

1 Introduction

If you want to find oil and gas accumulations, or produce them efficiently once found, then you need to understand subsurface geology. At its simplest, this means mapping subsurface structure to find structures where oil and gas may be trapped, or mapping faults that may be barriers to oil flow in a producing field. It would be good to have a map of the quality of the reservoir as well (e.g. its thickness and porosity), partly to estimate the volume of oil that may be present in a given trap, and partly to plan how best to get the oil or gas out of the ground. It would be better still to see where oil and gas are actually present in the subsurface, reducing the risk of drilling an unsuccessful exploration well, or even following the way that oil flows through the reservoir during production to make sure we don't leave any more of it than we can help behind in the ground. Ideally, we would like to get all this information cheaply, which in the offshore case means using as few boreholes as possible.

One traditional way of understanding the subsurface is from geological mapping at the surface. In many areas, however, structure and stratigraphy at depths of thousands of feet cannot be extrapolated from geological observation at the surface. Geological knowledge then depends on boreholes. They will give very detailed information at the points on the map where they are drilled. Interpolating between these control points, or extrapolating away from them into undrilled areas, is where geophysical methods can be most helpful.

Although some use has been made of gravity and magnetic observations, which respond to changes in rock density and magnetisation respectively, it is the seismic method that is by far the most widely used geophysical technique for subsurface mapping. The basic idea is very simple. Low-frequency sound waves are generated at the surface by a high-energy source (for example a small explosive charge). They travel down through the earth, and are reflected back from the tops and bases of layers of rock where there is a change in rock properties. The reflected sound travels back to the surface and is recorded by receivers resembling microphones. The time taken for the sound to travel from the source down to the reflecting interface and back to the surface tells us about the depth of the reflector, and the strength of the reflected signal tells us about the change of rock properties across the interface. This is similar to the way a ship's echo sounder can tell us the depth of water and whether the seabed is soft mud or hard rock.

Initially, seismic data were acquired along straight lines (2-D seismic); shooting a number of lines across an area gave us the data needed to make a map. Again, the process is analogous to making a bathymetric map from echo soundings along a number of ship tracks. More recently, it has been realised that there are big advantages to obtaining very closely spaced data, for example as a number of parallel straight lines very close together. Instead of having to interpolate between sparse 2-D lines, the result is very detailed information about the subsurface in a 3-D cube (x and y directions horizontally on the surface, z direction vertically downwards but in reflection time, not distance units). This is what is known as 3-D seismic.

This book is an introduction to the ways that 3-D seismic can be used to improve our understanding of the subsurface. There are several excellent texts that review the principles and practice of the seismic method in general (e.g. Sheriff & Geldart, 1995). Our intention is to concentrate on the distinctive features of 3-D seismic, and aspects that are no different from the corresponding 2-D case are therefore sketched in lightly. The aim of this first chapter is to outline why 3-D seismic data are technically superior to 2-D data. However, 3-D seismic data are expensive to acquire, so we look at the balance between better seismic quality and the cost of achieving it in different cases. The chapter continues with a roadmap of the technical material in the rest of the book, and concludes with notes on some important details of the conventions in use for displaying seismic and related data.

A complementary view of 3-D seismic interpretation, with excellent examples of colour displays, is provided by Brown (1999).

1.1   Seismic data

The simplest possible seismic measurement would be a 1-D point measurement with a single source (often referred to as a *shot*, from the days when explosive charges were the most usual sources) and receiver, both located in the same place. The results could be displayed as a seismic trace, which is just a graph of the signal amplitude against travel-time, conventionally displayed with the time axis pointing vertically downwards. Reflectors would be visible as trace excursions above the ambient noise level. Much more useful is a 2-D measurement, with sources and receivers positioned along a straight line on the surface. It would be possible to achieve this by repeating our 1-D measurement at a series of locations along the line. In practice, many receivers regularly spaced along the line are used, all recording the signal from a series of source points. In this case, we can extract all the traces that have the same midpoint of the source–receiver offset. This is a *common midpoint gather* (CMP). The traces within such a CMP gather can be added together (*stacked*) if the increase of travel-time with offset is first corrected for (*normal moveout (NMO) correction*). The details of this process are discussed in chapter 2.

The stacked trace is as it would be for a 1-D observation, with coincident source and receiver, but with much improved signal to noise ratio. These traces can then be displayed as a *seismic section*, in which each seismic trace is plotted vertically below the appropriate surface point of the corresponding 1-D observation. The trace spacing depends on the spacing of shots and receivers, but might be 12.5 or 25 m for a typical survey. The seismic section is to a first approximation a cross-section through the earth, though we need to note several limitations.

(1) The vertical axis is the time taken for seismic waves to travel to the reflector and back again (often called the *two-way time, TWT*), not depth.

(2) The actual reflection point in the subsurface is not necessarily vertically below the trace position, if the subsurface reflectors are dipping. We can try to reposition the reflection to the correct trace location so that the cross-section is closer to the real subsurface structure, but this is only in part possible for a 2-D line (see section 1.2).

(3) For a subsurface interface to generate a reflection, there has to be a change across it of a quantity called *acoustic impedance* (which is the product of density and seismic velocity in the layer concerned), so that not all interfaces of geological significance are necessarily visible on seismic data. The seismic velocity is the velocity with which seismic waves (see the glossary in Appendix 2) travel through the rock.

(4) The vertical resolution of the section, which is discussed further in chapter 4, is likely to be at best 5 ms. (TWT is usually expressed in milliseconds (ms): 1 ms = 1/1000 s.) Despite all this, the 2-D section gives considerable insight into the geometry of the subsurface.

Although not necessarily acquired in this way, a simple way of thinking of 3-D data is as a series of closely spaced parallel sections. The spacing between these sections might be the same 12.5 or 25 m as the typical trace spacing within each section. There are two benefits to be derived from the 3-D coverage:

(a) correcting for lateral shifts of reflection points in 3-D rather than 2-D produces a better image of the subsurface,

(b) the very dense data coverage makes it much easier and less ambiguous to follow structural or stratigraphic features across the survey area.

We shall discuss each of these in turn.

1.2 Migration of seismic data

The process of transforming the seismic section to allow for the fact that the reflection points are laterally shifted relative to the surface source/receiver locations is known as seismic *migration*. For a 2-D section, fig. 1.1 shows how the problem arises. We assume that the data as recorded have been transformed (as discussed above) to what would be observed in the *zero-offset* case, i.e. with source and receiver coincident and therefore no offset between them. For zero-offset, the reflected ray must retrace the

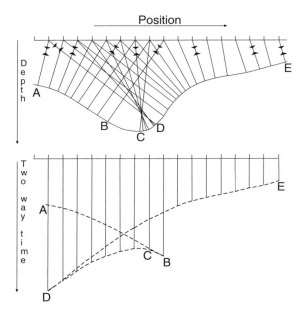

Fig. 1.1   Sketch of normal-incidence rays and resulting time section.

path of the incident ray to the reflector, so the angle of incidence at the reflecting horizon must be 90°. Not only are reflection points not directly below the surface point wherever this horizon is dipping, but for some surface locations there may be several different reflections from the horizon, and for other surface locations there may be no reflections received at all. The display produced by plotting seismic traces vertically below the surface points will, as sketched in the lower half of fig. 1.1, be hard to interpret in any detailed sense. This problem is solved by a processing step called *migration*, which repositions reflectors to their correct location in space. There are various ways of carrying this out in practice, but the basis of one method (*Kirchhoff summation*) is illustrated in fig. 1.2. This shows a point scatterer in a medium of uniform velocity; this reflector is to be thought of as a 'cat's eye' that reflects any incident ray directly back along the path by which it arrived. If a seismic line is shot above such a reflector, it appears on the resulting section as a hyperbolic event. This suggests a migration method as follows. To find the amplitude at a point A in the migrated section, the hyperbola corresponding to a point scatterer at A is superimposed on the section. Wherever it crosses a trace, the amplitude value is noted. The sum of these amplitudes gives the amplitude at A in the migrated section. Of course, not all the amplitude values in the summation truly relate to the scatterer at A; however, if there are enough traces, energy received from other scatterers will tend to cancel out, whereas energy truly radiated from A will add up in phase along the curve. (A more complete discussion shows that various corrections must be applied before the summation, as explained, for example, in Schneider, 1978.)

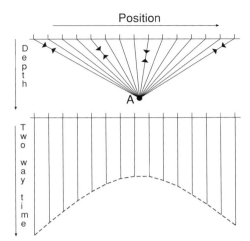

Fig. 1.2   Sketch of rays reflected from a point scatterer and resulting time section.

The snag with such a procedure is that it repositions data only within the seismic section. If data were acquired along a seismic line in the dip direction, this should work fairly well; if, however, we acquire data along a line in the strike direction, it will not give correct results. If we have a 2.5-D structure, i.e. a 3-D structure in which the dip section is the same at all points along the structure, then on the strike section all reflectors will be horizontal, and the migration process will not reposition them at all. After migration, dip and strike sections will therefore not tie at their intersection (fig. 1.3(a)). This makes interpretation of a close grid of 2-D lines over a complex structure very difficult to carry out, especially since in the real world the local dip and strike directions will change across the structure.

In general, some of the reflections on any seismic line will come from subsurface points that do not lie directly below the line, and migrating reflections as though they do belong in the vertical plane below the line will give misleading results. For example, fig. 1.3(b) shows a sketch map of a seismic line shot obliquely across a slope. The reflection points are located offline by an amount that varies with the local dip, but is typically 250 m. If we see some feature on this line that is important to precise placing of an exploration well (for example a small fault or an amplitude anomaly), we have to bear in mind that the feature is in reality some 250 m away from the seismic line that shows it. Of course, in such a simple case it would be fairly easy to allow for these shifts by interpreting a grid of 2-D lines. If, however, the structure is complex, perhaps with many small fault blocks each with a different dip on the target level, it becomes almost impossible to map the structure from such a grid.

Migration of a 3-D survey, on the other hand, gathers together energy in 3-D; Kirchhoff summation is across the surface of a hyperboloid rather than along a hyperbola (fig. 1.4). Migration of a trace in a 3-D survey gathers together all the reflected energy that belongs to it, from all other traces recorded over the whole (x, y) plane. This

(a)

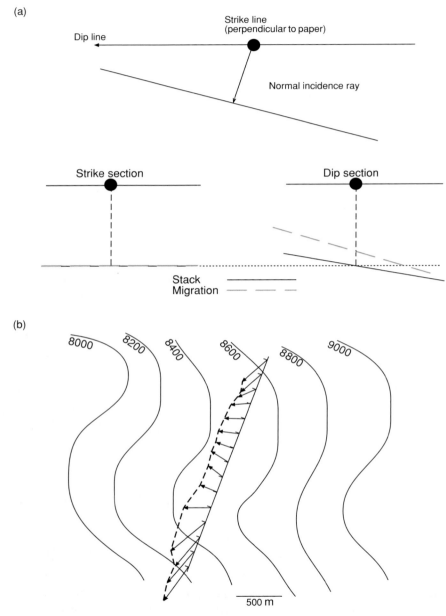

Fig. 1.3   (a) For a 2.5-D structure, dip and strike lines do not tie after migration; (b) map view of reflection points for a 2-D line (contours are depth in feet (ft)).

means that events are correctly positioned in the 3-D volume (provided that the migration process has been carried out with an accurate algorithm and choice of parameters, as discussed further in chapter 2). This is an enormous advance for mapping of complex areas; instead of a grid of lines that do not tie with one another, we have a

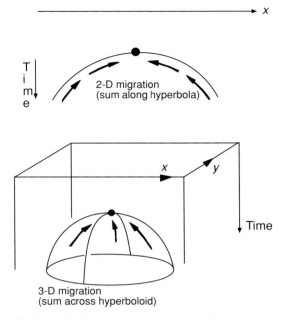

Fig. 1.4   Kirchhoff migration in 2-D and 3-D.

volume of trace data, from which sections can be chosen for display in any orientation we want. Furthermore, focussing of the data is also improved. For unmigrated data, the limiting horizontal resolution can be taken as the size of the Fresnel zone, an area surrounding any point on the reflector from which reflected energy arrives at the receiver more or less in phase and thus contributing to the signal at that reflection point. The radius f of this zone is given approximately by

$$f^2 = \frac{\lambda h}{2},$$

where λ is the dominant wavelength of the seismic disturbance and h is the depth of the reflector below the source–receiver point (see e.g. McQuillin *et al.*, 1984). This can amount to several hundred metres in a typical case. Migration collapses the Fresnel zones; 2-D migration collapses the zone only along the line direction, but 3-D migration collapses it in both inline and crossline directions, to a value approaching $\lambda/2$, which may amount to a few tens of metres. This will improve the detail that can be seen in the seismic image, though various factors will still limit the resolution that can be achieved in practice (see section 4.1).

1.3   Data density

When 3-D seismic first became available, it resulted in an immediate increase in the accuracy of subsurface structure maps. This was partly because of the improved imaging

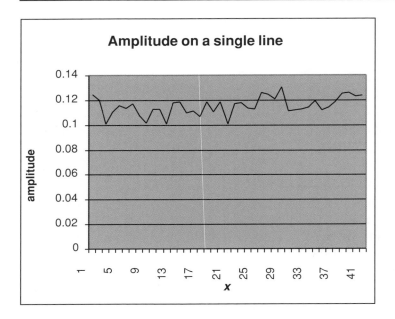

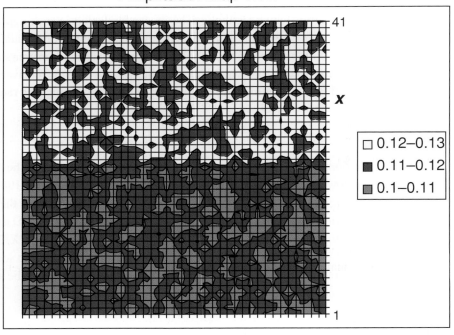

Fig. 1.5   Top: graph of amplitude versus position along a single line. Bottom: map view of amplitude variation across many similar parallel lines.

discussed in the last section, but also because of the sheer density of information available. Mapping complex structures from a grid of 2-D data is a subjective process; the interpreter has to make decisions about how to join up features seen on lines that might be a kilometre or more apart. This means that establishing the fault pattern in a complicated area will be time-consuming, and the resulting maps will often have significant uncertainties. 3-D data, with their dense grid of traces, allow features such as faults or stratigraphic terminations to be followed and mapped with much greater assurance (see section 3.2.2).

More recently, it has been realised that the density of coverage allows us to make more use of seismic attributes. This will be discussed in detail in chapter 5, but a typical example might be that we measure the amplitude of a seismic reflection at the top of a reservoir, which increases when hydrocarbons are present. Such an effect is often quite subtle, because the amplitude change may be small and almost lost in the noise in the data. Consistent changes across a 3-D dataset stand out from the noise much more clearly than changes along a 2-D line.

Figure 1.5 shows a synthetic example illustrating the power of seeing dense data in map view. At the top is a graph of amplitude along a single line; the left-hand half has a mean value of 0.11 and the right-hand half of 0.12, and uniformly distributed random noise with amplitude ± 0.01 has been added. Working from this graph alone, it would be hard to be certain that there is a higher average amplitude over the right-hand part, or to say where the change occurs. The lower part of fig. 1.5 shows a contour map of the amplitudes of 40 such lines, each with the amplitude step in the same place but a different pattern of random noise; the lines run from bottom to top of the area. It is immediately obvious that there is a step change in average amplitude and that it occurs halfway up the area. As we shall see in chapter 5, correlation of amplitude anomalies with structure can be a powerful test for hydrocarbon presence; this synthetic example shows why interpretation of amplitude anomalies is much more solidly founded on 3-D data than on a grid of 2-D data.

1.4   Uses of seismic data

Seismic data are used both in exploration for oil and gas and in the production phase. The type and quality of data gathered are determined by the balance between the cost of the seismic and the benefit to be gained from it. The general pattern is as follows.

(1) Early exploration. At this stage, knowledge will probably be very sketchy, with little or no well information. The presence of a sedimentary basin may be inferred from outcrop geology, or indirectly from geophysical methods such as gravity and magnetics that distinguish sedimentary rocks from metamorphic basement on the basis of their density or magnetic susceptibility (see e.g. Telford *et al.*, 1976).

At this stage even a small number of 2-D seismic profiles across the basin, perhaps tens of kilometres apart, will be very helpful in defining the general thickness of sediments and the overall structural style.

(2) Perhaps after some initial wells have been drilled in the basin with encouraging results, exploration moves on to a more detailed study, where the aim is to define and drill valid traps. More seismic data are needed at this stage, although the amount depends on the complexity of the structures. Simple anticlines may be adequately defined from a small number of 2-D profiles, but imaging of complex fault architectures will often be too poor on 2-D data for confident interpretation. If wells are fairly cheap and seismic data are expensive to acquire (as is often the case on land) it may be best to drill on the basis of a grid of 2-D lines. If wells are very expensive compared with seismic acquisition (the typical marine case), then it will already be worthwhile at this stage to use 3-D seismic to make sure that wells are correctly located within the defined traps. This might, for example, be a matter of drilling on the upthrown side of a fault, or in the correct location on a salt flank to intersect the pinchout of a prospective horizon. An example where 3-D seismic completely changed the structural map of a field is shown in fig. 1.6(a) (redrawn after Greenlee *et al.*, 1994). This is the Alabaster Field, located on a salt flank in the Gulf of Mexico. The first exploration well was drilled on the basis of the 2-D map and was abandoned as a dry hole, encountering salt at the anticipated pay horizon. The 3-D survey shows that this well was drilled just updip of the pinchout of the main pay interval. This is a case where seismic amplitudes are indicative of hydrocarbon presence and are much easier to map out on 3-D seismic.

(3) After a discovery has been made, the next step is to understand how big it is. This is the key to deciding whether development will be profitable. At this stage, appraisal wells are needed to verify hydrocarbon presence and investigate reservoir quality across the accumulation. Detailed seismic mapping may reduce the number of appraisal wells needed, which will have an important impact on the overall economics of the small developments typical of a mature hydrocarbon province. The next step will be to plan the development. An example of the impact of 3-D on development planning is shown in fig. 1.6(b) (redrawn after Demyttenaere *et al.*, 1993). This shows part of the Cormorant Field of the UK North Sea, where oil is trapped in Middle Jurassic sandstones in four separate westerly dipping fault blocks. The left-hand side of the figure shows the initial map of one of these fault blocks based on 2-D seismic data; the absence of internal structural complexity led to a development concept based on a row of crestal oil producers supported by downflank water injectors. The right-hand side of the figure shows the map of the fault block based on 3-D seismic; the compartmentalisation of the fault block led to a revised development plan with the aim of placing producer–injector pairs in each major compartment.

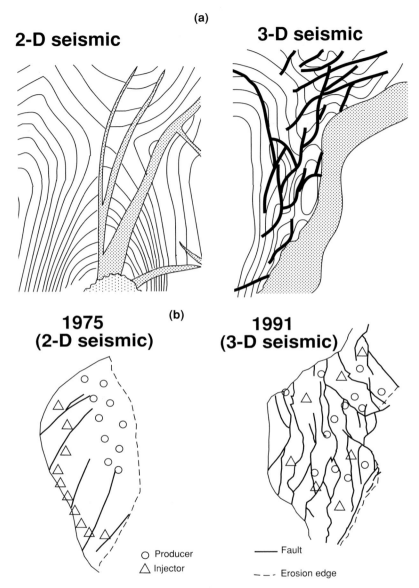

Fig. 1.6 Changes to maps owing to 3-D seismic: (a) Alabaster Field, Gulf of Mexico (redrawn after Greenlee *et al.*, 1994, with permission of the authors and the SEG); (b) Cormorant Field, UK North Sea (redrawn after Demyttenaere *et al.* (1993) with permission of the authors and the Geological Society of London).

(4) During field life, additional producers and injectors may be needed to optimise oil recovery. An accurate structural map will certainly be needed, and any information that can be gleaned from seismic on lateral variation of reservoir quality will be highly welcome. A further contribution from seismic is possible: we can sometimes

see how the distribution of hydrocarbons in the reservoir evolves during production by repeating the seismic survey after some production has taken place (4-D seismic, discussed in chapter 8). This will show where, for example, oil is not being swept towards the producer wells, perhaps because faults form a barrier to flow; additional wells can then be targeted on these pockets of bypassed oil. In this application, 3-D seismic is essential because the better focussing and denser data are needed to look for subtle clues to reservoir quality and hydrocarbon presence.

The decision on whether or when to shoot 3-D seismic is essentially an economic one. Does the value of the subsurface information obtained justify the cost? This issue has been discussed by Aylor (1995), who collated data on 115 3-D surveys. At the time, the average cost for a proprietary marine survey was US $4.2 million, and for a land survey it was US $1.2 million. The average 3-D development survey resulted in the identification of six previously unknown high-quality drilling locations. It also separated good from bad locations: before 3-D the average probability of success (POS) of a well was 57%, whereas after 3-D the locations fell into two groups, with 70% of locations having an increased POS of 75%, and the remaining 30% of locations having a much reduced POS of only 17%. 3-D seismic was also very effective at targeting sweet spots in the reservoir: initial production rates per well averaged 565 barrels per day (b/d) without 3-D and 2574 b/d with it. Using this information together with information on direct 3-D survey costs (for acquisition, processing and interpretation), and the indirect costs due to the delay in development while the 3-D survey was being acquired and worked, it was calculated that the average 3-D survey added US $14.2 million in value, most of which came from the addition of previously unrecognised drilling locations and the higher initial production rates. Results from this limited database thus indicate the positive value of using 3-D seismic. Studies of a larger database would be instructive, but unfortunately industry-wide information on 3-D seismic costs and benefits is elusive (Nestvold & Jack, 1995).

However, the oil industry as a whole is convinced of the value of 3-D survey, as can be seen from the growth of 3-D seismic acquisition worldwide. According to a survey by IHS Energy Group (summarised in *First Break*, **19**, 447–8 (2001)), onshore annual 3-D acquisition increased from 11 000 sq km in 1991 to 30 000 sq km in 2000, while annual offshore 3-D acquisition rose from 15 000 sq km to 290 000 sq km. Over the same period, 2-D acquisition fell from 260 000 line km to 110 000 km onshore, and from 1 300 000 km to 840 000 km offshore. The striking increase in offshore 3-D coverage is no doubt due to the efficiency of marine acquisition and resulting low cost per square kilometre. It may also reflect the high fixed costs of marine survey. Almost all modern marine seismic is shot by specialist contractors, who need to keep their boats working continuously; this results in a mix of commercial arrangements, including surveys shot exclusively for one oil company, surveys shot for a group of companies, surveys shot at the contractor's risk with the intention of selling the final processed data on the open market, and various hybrids between them.

1.5   Road map

Chapter 2 is devoted to explaining how 3-D seismic data are acquired and processed. The interpreter needs to have at least an outline knowledge of these topics, for two reasons. One of them is the need to understand what the limitations of the data are. Often, the interpreter is struggling to get as much information as possible out of a seismic dataset, and has to decide how far his conclusions are robust, or whether there is a chance that he is being misled by noise. The other reason is that the interpreter will be asked, when his best efforts still leave him unsure about the subsurface picture, what can be done to improve the data. He will then sometimes find himself in a dialogue with acquisition and processing experts, and need to speak their language. Chapter 2 aims at equipping him to do this. Although most space is given to the specific issues that arise for 3-D, the methods that are no different from the 2-D case have been sketched in to give a reasonably complete account.

Chapters 3 and 4 describe the basic interpretation process. The distinction between structural and geological interpretation is an artificial one, in the sense that both are going on simultaneously as the interpreter works through his data. However, many interpreters spend much of their time making structural maps or planning well trajectories. Therefore, the basic mechanics of workstation interpretation are covered at some length in chapter 3. Chapter 4 considers some of the ways that 3-D seismic can lead to enhanced geological understanding, and what some of the problems are, especially because of the limited resolution of seismic data.

The availability of dense grids of data has revolutionised our ability to make useful inferences from measuring seismic attributes, such as the detailed study of amplitudes of individual seismic loops. This topic is therefore covered in detail in chapter 5. Inversion of seismic data to acoustic impedance is covered in chapter 6; this is an old idea that has become much more useful with the availability of high-quality dense 3-D datasets. It converts the standard seismic section, which emphasises the layer boundaries where reflections occur, into a form that emphasises the properties of individual layers. This can then be a starting-point for discussions with other disciplines, for example the reservoir engineer.

An area of rapid progress at present is the use of more powerful computer workstations to give the interpreter a better appreciation of the 3-D nature of the subsurface, viewing 3-D bodies directly rather than just seeing 2-D sections through them. This is explained in chapter 7.

There is increasing interest in using repeated surveys over producing fields to follow changes in seismic response produced by changes in porefill (e.g. development of a gas cap); this is another old idea that has become more feasible with the availability of high-quality 3-D surveys. Such surveys, usually called 'time-lapse' or '4-D' seismic, are discussed in chapter 8. Appendix 1 contains a brief summary of the hardware and software issues involved in managing interpretation workstations in practice, and finally Appendix 2 contains a glossary of technical terms. This is not intended to be exhaustive; a definitive dictionary of geophysical terms has been compiled by Sheriff (1991).

1.6   Conventions: seismic display, units

There are two topics to mention here that may cause confusion to the unwary reader: display polarity and systems of units. Display polarity is the more important of these; arguing about polarity wastes large amounts of interpreter time. The problem is this: if we have an interface at which impedance increases downwards, when we make a wiggle display of a seismic trace with the time axis vertical, does such an interface give rise to a deflection to the left (a trough) or to the right (a peak)? Classically on paper sections, the peaks were shaded to produce a display in which the peaks were black and the troughs appeared white, and a similar convention is often used for workstation displays so that peaks are black or blue and troughs are red or white. In principle, polarity is fixed at the time of recording the data, and preserved throughout the processing sequence. Many modern datasets are transformed to zero-phase so that a single interface is represented by a single loop with some lower-amplitude wiggles on either side of it. Processors often describe the polarity of the final data as 'SEG normal' or 'SEG reverse'. This refers to a convention promulgated by the Society of Exploration Geophysicists (SEG), according to which SEG normal would correspond to an increase in impedance downwards being represented by a peak. The reverse convention (SEG reverse) is commonly employed in some hydrocarbon provinces, e.g. the UK North Sea. Unfortunately, it is quite possible for mistakes in acquisition or processing to result in final displays with polarity opposite to the processors' stated convention. The interpreter needs to check for himself what the polarity of a given dataset really is. It might be thought that this can easily be done by comparing the data with well synthetics (section 3.1). However, many seismic sections show long intervals of wiggles of about the same amplitude and frequency, and over such an interval it may be easy to establish plausible ties using either polarity convention, if bulk shifts are allowed; such shifts are almost always present in real data owing to various limitations in processing. A good check is to find an isolated interface with a large and sharp impedance change across it, which will give rise to a strong and isolated seismic loop; inspection of this will reveal the polarity of the data. In the marine case, it is tempting to use the seabed for this purpose, but care is needed because the true seabed may have been removed by the application of a trace mute during processing. In view of the potential for confusion, it is good practice to state polarity explicitly in discussions of seismic data, with some phrase such as: 'increase of impedance downwards gives rise to a red (or blue, or black, etc.) loop'. This is a cumbersome convention, but is at least clear. In some cases, where conversion to zero-phase has not been carried out or has been unsuccessful, a single isolated impedance interface may give rise to a complicated reflection signal, with several loops of roughly the same amplitude. In this case, polarity is not a meaningful idea, and a sketch of the response of an isolated interface should accompany seismic displays. A more detailed discussion of these issues has been given by Simm & White (2002).

There is no uniform convention in the industry regarding units of distance. Both feet/inches and kilometres/metres/centimetres units are commonly employed, and are often freely mixed (e.g. horizontal distances in metres and vertical distances in feet). This is easy to cope with using the conversions in section 1.7. In this book, both systems are used depending on the source of data under discussion. In the real world, units are almost invariably annotated on displays, so confusion should be minimal. Much more confusion is generated by inadequately documented displays of well data; depths may be as measured (measured depth, the distance along hole from a fixed reference point, e.g. the derrick floor of the drilling rig), or relative to a geographical datum (usually sea-level for marine data), or may have been corrected for well deviation to give vertical depths (again, relative to derrick floor, sea-level, etc.). Close inspection of displays of well logs is often needed to establish what the depth reference actually is. (A similar problem arises with onshore seismic data, where zero time will usually correspond to a datum plane at some particular elevation, which may not however be documented in a way that is easily retrievable.) Another possible source of confusion in horizontal positioning of well and seismic data arises from the use of different map projection systems. Many different systems are in use, and even within a given projection system there are different possible choices for projection parameters. This can easily cause problems in relating wells to seismic survey data. Since new well locations are usually chosen from a seismic survey grid, at the intersection of a particular inline and crossline, it is obviously critical to be able to translate this intersection into a point on the ground where the rig will actually be placed. One of the problems of 3-D seismic is that interpretation is often carried out on a more or less self-contained workstation volume; such volumes often have a complicated history of reprocessing by different people at different times, and it may not be easy to check whether the real-world coordinates assigned to this volume are correct. If there is any doubt at all about the coordinate systems being used, the services of a specialised surveyor are needed.

1.7   Unit conversions

Conversions are stated to four significant figures where not exact.

Length: 1 inch = 2.540 cm
         1 foot = 12 inches = 30.48 cm
         1 metre = 100 cm = 3.281 ft

Density: $1\,\mathrm{g/cm^3} = 1000\,\mathrm{kg/m^3} = 62.43\,\mathrm{lb/ft^3}$
Volume: $1\,\text{litre} = 1000\,\mathrm{cm^3} = 0.035\,31\,\mathrm{ft^3}$
           $1\,\text{barrel (bbl)} = 0.1590\,\mathrm{m^3}$
           $1\,\mathrm{m^3} = 6.290\,\text{bbls}$

References

Aylor, W. K. (1995). Business performance and value of exploitation 3-D seismic. *The Leading Edge*, **14**, 797–801.

Brown, A. R. (1999). *Interpretation of Three-dimensional Seismic Data* (5th edn). American Association of Petroleum Geologists, Tulsa.

Demyttenaere, R. R. A., Sluijk, A. H. & Bentley, M. R. (1993). A fundamental reappraisal of the structure of the Cormorant Field and its impact on field development strategy. In: *Petroleum Geology of Northwest Europe: Proceedings of the 4th Conference*, J. R. Parker (ed.), pp. 1151–7, Geological Society, London.

Greenlee, S. M., Gaskins, G. M. & Johnson, M. G. (1994). 3-D seismic benefits from exploration through development: an Exxon perspective. *The Leading Edge*, **13**, 730–4.

McQuillin, R., Bacon, M. & Barclay, W. (1984). *An Introduction to Seismic Interpretation* (2nd edn). Graham & Trotman Ltd, London.

Nestvold, E. O. & Jack, I. (1995). Looking ahead in marine and land geophysics. *The Leading Edge*, **14**, 1061–7.

Schneider, W. A. (1978). Integral formulation for migration in two dimensions and three dimensions. *Geophysics*, **43**, 49–76.

Sheriff, R. E. (1991). *Encyclopedic Dictionary of Exploration Geophysics* (3rd edn). Society of Exploration Geophysicists, Tulsa.

Sheriff, R. E. & Geldart, L. P. (1995). *Exploration Seismology*. Cambridge University Press, Cambridge, UK.

Simm, R. & White, R. (2002). Phase, polarity and the interpreter's wavelet. *First Break*, **20**, 277–81.

Telford, W. M., Geldart, L. P., Sheriff, R. E. & Keys, D. A. (1976). *Applied Geophysics*. Cambridge University Press, Cambridge, UK.

2 3-D seismic data acquisition and processing

The aim of seismic data acquisition and processing is to deliver products that mimic cross-sections through the earth. In order to do this, the correct amount and types of data must be acquired, and processing applied to remove unwanted energy (such as multiples), and to place the required events in the correct location. At the same time, a balance needs to be struck between cost and timeliness of data, while attaining also the important objectives of safe operations and doing no harm to the environment.

It is not the aim of this chapter to give a full account of seismic acquisition and processing; rather we aim to concentrate on those aspects that are specific to 3-D operations or are recent innovations. For those who require more information on energy sources, instrumentation, receivers and general acquisition theory there are a number of detailed references such as Stone (1994) and Evans (1997). In addition there are good introductions in some of the more general texts such as Sheriff & Geldart (1995) or McQuillin et al. (1984).

The vast bulk of seismic data currently acquired is 3-D, owing to the tremendous advantages in terms of interpretability discussed in chapter 1. Today it is unusual for the major oil companies to drill exploration wells prior to a 3-D survey being shot, processed and interpreted. Surveys range in size from a few tens of square kilometres for field development to several thousand square kilometres for exploration purposes in frontier basins. Land 3-Ds are less common than marine, partly because of their higher cost, but also because land well costs are relatively low.

In order to achieve the aims outlined above, surveys need to be planned to cover adequately the area of interest, taking into account data repositioning due to migration. To achieve this, the actual recorded data must cover an area that is larger than the target area by a migration aperture (fig. 2.1 and later in this chapter). In addition, the trace spacing needs to be small enough in all directions to avoid data aliasing. Ideally, subsurface coverage should be uniform with a consistency between the contribution from different offsets (the distance between the source and receiver) and azimuths (the direction between the source and the receiver). Budget, access, water currents or timing issues may mean one or more of these guidelines will have to be sacrificed and a balance struck between operational expediency and data quality.

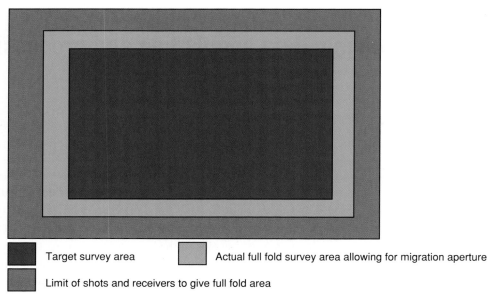

Target survey area   Actual full fold survey area allowing for migration aperture

Limit of shots and receivers to give full fold area

Fig. 2.1   Relationship between target area and acquisition area.

Modern powerful computers enable data processing to begin in the field or on the boat shortly after the acquisition has started, leading to rapid delivery of products even for the largest surveys. This is important not just for the financial implications of improved turnaround time, but also because decisions on data quality, such as the effects of bad weather and increased swell noise, can be made by examination of real data quality. Only the most time-consuming processes such as pre-stack 3-D migration need dedicated processing centres with large powerful computers.

2.1   Marine 3-D data acquisition

In general, 3-D marine data acquisition is simpler and faster than land acquisition since in all but the most heavily developed offshore areas there are few obstacles, leading to routine and rapid data gathering. In standard marine acquisition, a purpose-built boat (fig. 2.2) is used to tow one or more energy sources and one or more cables containing (pressure sensitive) receivers to record the reflections from the underlying rocks. At present, the source is nearly always an array of air guns tuned to give an energy pulse of short duration with directivity characteristics that concentrate the energy vertically downwards. In the past, other sources such as water and steam guns were used, and these may be encountered on older 3-D surveys. Evans (1997) gives a good description of the workings of an air gun; briefly, the expansion and collapse of the air bubble in the water acts as an acoustic source that sends sound waves through the water and into the rock layers below the seabed. At changes in the rock acoustic

Fig. 2.2   Examples of modern marine 3-D seismic vessels.

impedance, part of the sound wave is reflected back to the surface where it is captured by the receivers and transmitted to the boat for further processing or writing to tape for storage (fig. 2.3).

In the early days (mid-1980s) of 3-D data acquisition the boats were not powerful enough to tow more than one cable and one set of guns, so the 3-D acquisition geometry was just a series of closely spaced 2-D lines (fig. 2.4). Owing to the high operating expense of this design, the surveys tended to be rather small and used only over developing fields. To reduce time wasted during operations several novel geometries were tried including circle and spiral shooting (fig. 2.5). These had varying degrees of success but certainly increased the difficulty of data processing. They do not lend themselves to modern multi-cable surveys. The benefit of the much clearer subsurface picture obtained from 3-D data led to a growing demand for such surveys and rapid advances in technology. Today, specially designed seismic vessels are powerful and sophisticated enough to tow multiple cables and deploy two or more gun arrays that are fired alternately. This allows multiple subsurface lines to be collected for each pass of the boat, significantly increasing the efficiency and reducing data acquisition costs. Figure 2.6 shows a typical mid-1990s layout of four cables and two gun arrays, giving

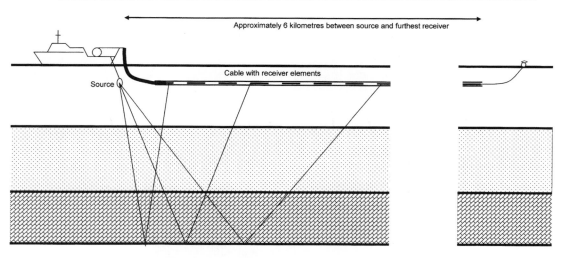

Approximately 6 kilometres between source and furthest receiver

Cable with receiver elements

Source

Example Reflection Points between a single shot and receivers

Fig. 2.3   Basics of marine acquisition. The boat travels through the water and every few metres fires the source which emits a sound wave into the water. This travels through the water and into the rock layers. At changes in the acoustic properties of the rock, as generally occur wherever the lithology changes, part of the sound wave is reflected back. The reflection travels up to the surface where it is captured by receivers within a long cable towed behind the boat. The receivers transmit the recorded signal back to the boat where it is stored on tape and may be partly processed.

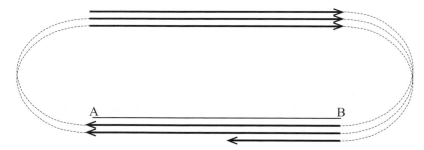

A   B

Fig. 2.4   Basic 3-D acquisition. After shooting a line, the boat turns with a relatively large radius before shooting a line in the opposite direction. The boat then turns again and shoots a line adjacent to the original line. This is repeated several times until finally the line AB is shot. The full survey may contain several repetitions of this basic design.

eight subsurface lines for each pass of the boat. The guns fire alternately; each gun generates as many subsurface lines as there are cables being towed. The gun separation is such that the lines recorded by one gun are interleaved with the lines shot by the second gun. Usually the separation between lines is between 25 and 37.5 m. It is diffi-cult to reduce this further with a single pass of the boat without an unacceptable risk of cables becoming entangled, but occasionally surveys are shot with 12.5 m line spacing

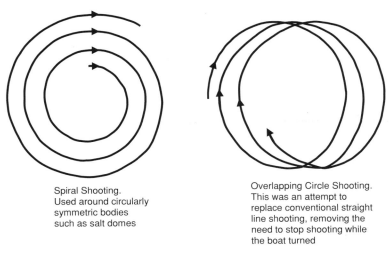

Spiral Shooting.
Used around circularly
symmetric bodies
such as salt domes

Overlapping Circle Shooting.
This was an attempt to
replace conventional straight
line shooting, removing the
need to stop shooting while
the boat turned

Fig. 2.5   Early non-standard 3-D marine acquisition techniques designed for efficient operations.

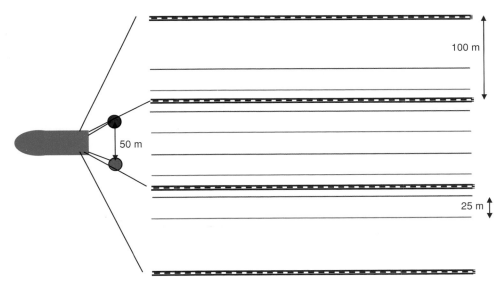

Fig. 2.6   The basics of multi-source, multi-streamer acquisition. As the boat sails along, the blue and the red gun array fire alternately. When the red guns fire, the four red subsurface lines are acquired. When the blue guns fire, the blue subsurface lines are recorded. The configuration shown gives eight subsurface lines per sail-line of the boat, 25 m apart from each other.

and modern steerable cables may make this more common in the future. The separation between traces recorded along a line is controlled by the receiver spacing and is usually between 6.25 and 12.5 m. As with the original marine survey design (fig. 2.4), a number of lines are shot together with the boat travelling in one direction, followed by a similar section with the boat travelling in the opposite direction. Occasionally, the

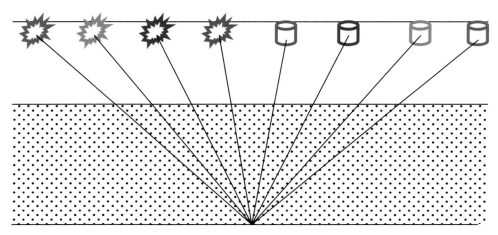

Fig. 2.7 Generation of multi-fold coverage during seismic surveys. As the survey proceeds, different combinations of shot and receiver record the same reflection point. The difference between the arrival times of the reflection from shot–receiver pairs with increasing separation allows subsurface velocities to be estimated. The addition of different traces with the same reflection point improves the signal to noise ratio of the final section.

areas of the survey where the boat changed shooting direction may lead to stripes in the data. This is caused by the long time interval between these adjacent lines being shot, so that there may be changes in tide, wave, currents or even water temperature (and hence water velocity) between the two lines. Other patterns may appear in the data, for instance owing to failure of part of the source gun array. Such striping may be difficult to remove by later data processing.

As with 2-D operations, the same subsurface location is recorded many times by traces having increasing separation between the source and receiver position (fig. 2.7). During processing, this multiplicity of data is used to increase the signal to noise ratio of the final stack, to pick subsurface velocities and to discriminate between different recorded events (such as primary and multiple reflections).

Modern boats are capable of towing as many as 12 cables each typically between 4 and 8 km in length, though it is unusual to see more than 10 actually deployed. This large increase of data gathered with each pass of the boat means that, despite the high cost of building and maintaining these vessels, their operational efficiency permits large exploration surveys, covering thousands of square kilometres, to be routinely acquired in a matter of weeks. At the time of writing, boats acquire up to 40 km^2 of data daily at costs of around US \$5000 per square kilometre. As a result of the high initial cost of marine vessels and the need to have near-continuous operations to be financially effective, almost all marine 3-D seismic boats are owned by specialised service companies and not by energy companies.

In order to be able to tow the cables through the water off to the side of the boat, large devices called paravanes are used to position the head of the outermost cables to the

Fig. 2.8 A paravane used to control the position of the head of the receiver cable.

required locations. Figure 2.8 shows the immense size of these devices. Towing several closely spaced multi-kilometre-long cables requires the boat to maintain control over their positioning at all times. Having cables cross over and wrap around each other is an expensive and dangerous situation and is to be avoided if at all possible. To keep control, the boat needs to be continuously moving through the water at a speed of at least 3 knots, and normal operations are carried out at a speed of around 4 knots. It also means that the seismic vessel requires a large turning area and needs to remain clear of other shipping in the area. Often, smaller boats known as picket vessels are used during the seismic survey to warn other boats to keep their distance while operations are underway. The actual length of the receiver cables depends on the depth to the target and also on the expected velocity profile in the earth. A deeper target requires a longer cable in order to pick velocities accurately and remove multiples. A good rule of thumb is that the cable length should be at least as long as the depth to the main target level, although longer cables may be required under special circumstances. The ends of the cables are attached to a tail buoy that contains a radar reflector to warn shipping of its presence and to enable the position of the cables to be monitored and recorded.

At the end of the required surface line the boat needs as much as 2 h to turn safely ready to shoot the next swathe of lines. This is non-productive time and so data are usually acquired with the longest axis of the survey area as the shooting direction for the sake of efficient operations. Other considerations such as dominant dip of the subsurface and current strength and direction are also taken into account when designing marine seismic operations. Data are usually shot in the dominant subsurface dip direction (fig. 2.9) as this is usually the direction with the smallest distance between recorded traces. There may be circumstances such as those illustrated in fig. 2.10 where strike or another shooting direction is preferred. For some targets, such as those under irregular salt bodies exhibiting large velocity contrast with the surrounding sediments, there may be no single optimum acquisition direction; surveys may have to be shot in several different directions to achieve good results. Currently this is rare, but is likely to become more common as companies experiment with multi-direction shooting. In an area with strong

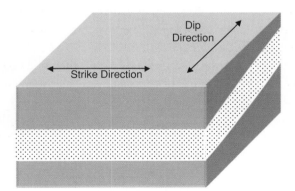

Fig. 2.9 Dip and strike directions. Over the years there have been several debates on the relative merits of shooting in the dip and strike directions. The argument generally hinges on velocity complexity. Velocities are complex and affected by dip in the dip direction, whereas they are not affected by dip in the strike direction. However, processing can unravel this complexity. In practice, most acquisition is in the dip direction, largely because the direction of shooting is usually the most densely sampled and therefore less prone to data aliasing.

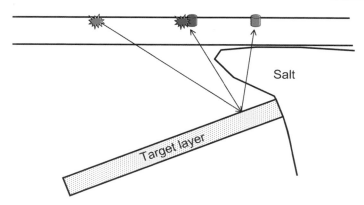

Fig. 2.10   Strike shooting over a salt dome. Dip shooting (in the plane of the section) would see complex ray-paths due to the presence of the salt, while strike shooting (in and out of the page) would be largely unaffected by the salt.

currents, it may be difficult for the boat to travel in or against the direction of the current and maintain good control over the cables and the required shot point spacing, leaving little choice on shooting direction.

Modern seismic boats are designed to tow large numbers and lengths of receiver cables through the water while remaining acoustically quiet so as to avoid unwanted noise on the recorded data. Special expansion joints are used in the cables to ensure that the cables themselves do not transmit noise along their length. The streamers have neutral buoyancy in the water so that they are easy to tow at a fixed depth (usually between 3 and 11 m depending on the seismic resolution required and the sea state) below the sea-surface. This avoids near-surface noise problems caused by the action of waves. In the past cables were filled with oil-based fluids to give them neutral buoyancy, but solid streamers are becoming more common as they are both environmentally friendlier and acoustically quieter. Special winged devices called 'birds' are attached to the cable to keep it at the required depth. These are remotely controlled by an on-board computer and permit changes in receiver depth (for instance in response to changing weather conditions) during operations without any need to bring the cables back on board.

To be able to shoot and process data accurately from a 3-D survey the boats have sophisticated navigation systems. During operations the boat and cables are subject to currents and winds, which will often cause the cables to be pushed away from the simple line-astern configuration, a situation known as 'feathering' (fig. 2.11). This means that the positions of shots and receivers must be continuously recorded, and this information needs to be stored together with the trace data. Today the entire world has 24 h satellite coverage that allows the use of GPS satellite positioning of the vessel to an accuracy within a metre. The streamers have a series of compasses and transducers along their length. The transducers send out signals whose travel times

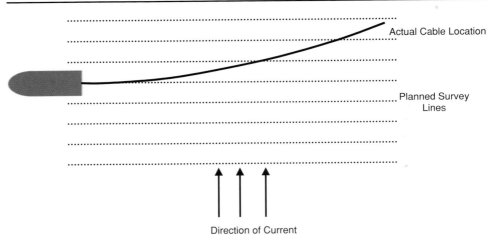

Fig. 2.11   Cable feathering due to action of current.

between each other and fixed receivers on the boat give a wealth of information on their position relative to each other and the boat. This information is used by powerful on-board computers to provide accurate measurements of absolute positions of every shot and receiver throughout the survey. This information is later merged with the seismic trace data during processing. On board, the navigation is also used to check that the survey has actually been shot within the requirements defined by the client. In many surveys it is necessary to re-shoot some lines because the original lines were not positioned correctly due to wind/wave/current effects. This re-shooting, known as infill, is an extra expense and every effort is made to keep it to a minimum by accurate steering of the boat. On-board computer systems predict the heading required to allow the boat to steer along the required trajectory taking into account the current conditions, and ensure that the shots are automatically fired at the correct locations. Despite all this, the effects of currents and winds usually mean that some infill is required.

In areas where there is already infrastructure (platform and production facilities) in place it is generally not possible to shoot the entire survey with the standard marine configuration. A safe distance must be kept between the boat and any obstacles. In such a case it is common to use techniques such as undershooting, where a source boat on one side of the obstacle shoots into a receiver boat on the other side (see fig. 2.12). Care must be taken during interpretation that any differences in the data in the undershot area are not a result of the different acquisition parameters.

2.2   Marine shear wave acquisition

Most marine acquisition to date has been concerned with compressional (P) wave re-flections from a P-wave source, since fluids (including sea-water) do not permit the

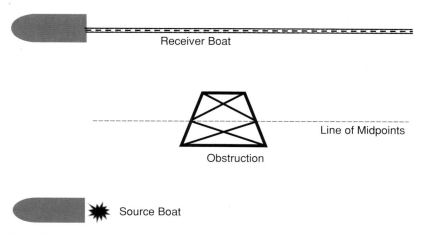

Fig. 2.12   Principles of undershooting. Boats need to keep a safe distance from any obstacles and therefore cannot acquire data close to rigs, etc. In order to record reflections from underneath such obstacles, undershooting is used. This is achieved using two boats, a source boat and a receiver boat. The source boat and receiver boat stay on either side of the obstacle with the source boat firing shots to be recorded by the second boat. The reflection point lies halfway between source and receiver, giving lines of reflected data from underneath the obstacle. It is still not possible to acquire some of the short offset data, but the result is much better than having a no-data zone within the survey. Since the source boat is not towing a large cable it is able to approach closer to the obstruction.

transmission of shear (S) waves. Recently (Garotta (1999) and Tatham & McCormack (1991)) there has been an interest in the use of marine shear waves, because they have advantages in areas where gas clouds obscure deeper reflectivity and also because they may produce stronger reflections in areas of low P-wave acoustic contrasts. Another proposed use is in areas with very high velocity contrasts, such as in imaging through salt or basalt layers; currently there are very few documented successes. In such cases, the S-wave velocity in the high-velocity zone can be close to the P-wave velocity in the surrounding sediments, giving a simpler ray-path for the converted ray and the possibility of increasing the angular coverage at depth (fig. 2.13). Marine shear wave exploration requires conversion of the P-wave input to a shear wave in the subsurface, normally at the target horizon, together with ocean-bottom receivers to allow the reflected shear wave to be recorded (fig. 2.14). Standard marine acquisition contains some data that have converted from P to S and back to P again. However, it is low in amplitude due to the double conversion, and the similarity in terms of moveout to multiples means that its identification and separation is extremely difficult.

The ocean-bottom receiver contains four phones, one pressure-sensitive hydrophone as in normal marine seismic recording and three mutually perpendicular velocity-sensitive geophones. This is often referred to as *four-component* recording, or 4C for short. A variety of different recording configurations have been developed over

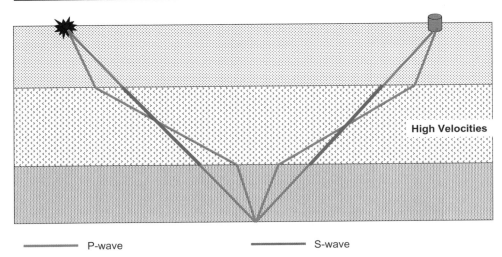

P-wave   S-wave

Fig. 2.13 The simple geometry of converted modes through high-velocity layers. The diagram shows the potential benefit of a double converted ray-path. The high-velocity layer leads to a large amount of ray-bending for the P-wave ray-path and a lack of angular coverage at depth. The ray-path that contains the P to S conversion through the high-velocity layer is much simpler and gives larger angular coverage at the target. In practice, it is very difficult to separate out such double conversions because of their low amplitudes and similarity in velocity to multiples.

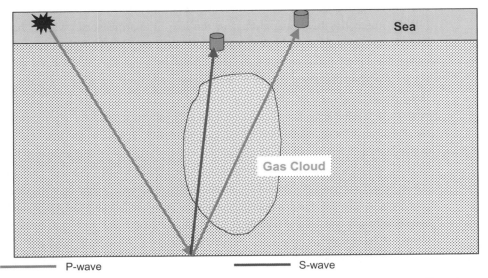

P-wave   S-wave

Fig. 2.14 Conventional marine shear wave geometry. Gas-saturated sediments are very disruptive to the passage of P-wave energy. The P-wave reflection will therefore be weak and highly scattered. On the other hand, the reflected S-waves are not affected by the presence of gas and travel through undisturbed. The shear wave data are recorded at the seabed to avoid energy loss at a second conversion back to P-wave.

the years. Multiple deployments of single phones that sink to or are planted in the seabed and later recovered is one technique. More popular are cables that are laid on or dragged across the seabed during acquisition. A recent innovation is cables that are permanently buried over a producing field and are hooked up to a recording vessel whenever a survey is required. Although the initial survey using this last technique is very expensive, any repeat survey is relatively cheap since the boat is not required to lift and drag the cables. In operations using cables, two vessels are needed though each can be substantially smaller and less powerful than a standard marine 3-D boat. One boat is attached to umbilicals from the cables and is the recording boat, remaining stationary while a second boat tows the source and sails around the survey area shooting a dense pattern of shots. In this way the acquisition becomes more similar to land operations, where it is also common to shoot a series of shots from different locations into static receiver cables. In all cases the separation between lines of receivers is much greater than with standard marine techniques, fold being built by a dense coverage of shots. Figure 2.15 shows two examples of marine four-component shooting geometries. The source used is generally a standard marine source, although occasionally with some tuning of the gun array to give a signature that is more uniform over a larger range of angles than the more directional source used in standard 3-D seismic surveys.

Both acquisition costs and processing time are greater for converted wave data, so it is seldom used in areas with a good P-wave image. Indeed, the amplitude versus offset information in conventional P-wave data contains information on the shear wave contrasts (see chapter 5). However, the technique has proved useful in areas where the P-wave image is obscured by gas clouds.

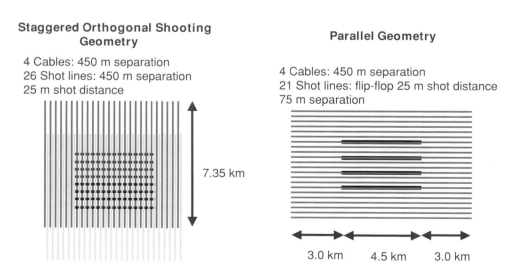

Fig. 2.15 Typical marine four-component geometries (not to scale).

Fig. 2.16   Land acquisition examples.

2.3   3-D land acquisition

There are a number of major differences between marine and land acquisition. In the latter, the shots and receivers are decoupled, the survey area has non-uniform topography, and obstacles and hazards are more common. Land operations occur in a variety of areas, from baking deserts to frozen wastes, and from jungles to cities (fig. 2.16). Each survey needs to be designed specifically for the terrain and region covered so as to ensure good data quality and optimised turnaround time, while being non-intrusive to the environment and ensuring the safety of the seismic operation's crew and local inhabitants.

The favoured land source is the vibrator since it is the most efficient to operate, has little environmental impact, and has better control over the source characteristics than the main explosive-based alternatives. The vibrator is a large truck that contains a mechanical device for shaking the ground through a controlled set of frequencies. Full details are given in the acquisition references included at the end of this chapter.

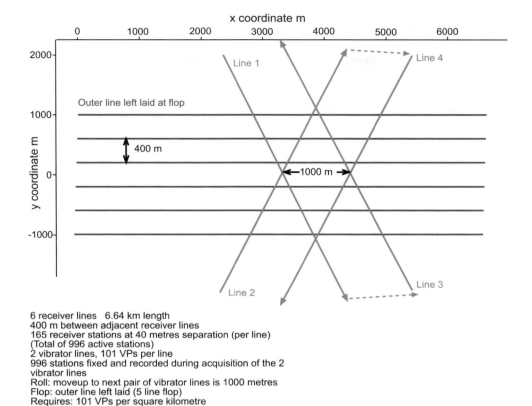

6 receiver lines   6.64 km length
400 m between adjacent receiver lines
165 receiver stations at 40 metres separation (per line)
(Total of 996 active stations)
2 vibrator lines, 101 VPs per line
996 stations fixed and recorded during acquisition of the 2
vibrator lines
Roll: moveup to next pair of vibrator lines is 1000 metres
Flop: outer line left laid (5 line flop)
Requires: 101 VPs per square kilometre
75 receiver stations per square kilometre

Fig. 2.17   Land acquisition. Basic X design.

Several vibrator trucks are often used together to increase the energy of the source, and cycled several times at the same location so as to be able to sum the results of different shots and increase the signal strength relative to noise in the records. After shooting is completed at one shot location (often called vibrator point or VP for short), the trucks drive to the next location. In areas of easy access and few obstacles this is a relatively efficient process and it is possible to shoot as many as 800 VPs per day. This is, however, still far below the efficiency achieved in marine work. Source and receiver positions are surveyed (generally by Global Positioning System, GPS) and are marked in advance by flags so as to allow rapid movement to the next VP. Once a section has been shot, the cable is picked up by the field crew and added further along the line ready for later shooting, so as to allow continuous operations during daylight hours.

Many different layouts of shots and receivers are available for land operations in areas without obstructions, and the skill of the survey designer is to ensure the most efficient arrangement that provides the required data quality. Figures 2.17–2.19 show how the

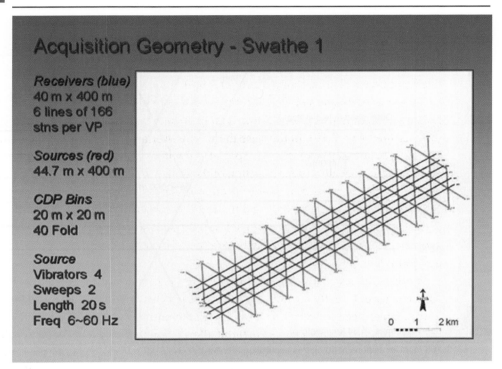

Fig. 2.18   Land acquisition. Completion of a swathe of data by repetition of the basic X design.

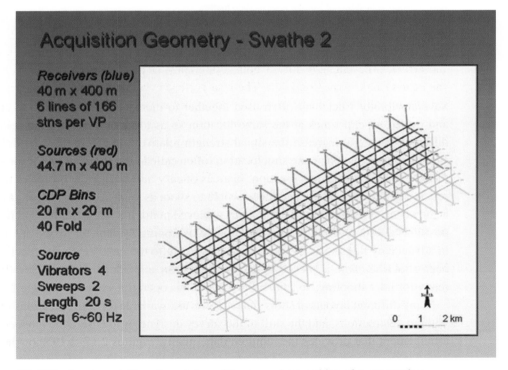

Fig. 2.19   Land acquisition. Completion of the survey by repetition of many swathes.

total survey is generally built up from continuous repetition of the basic acquisition pattern or footprint. In fig. 2.17 we see six receiver lines with shot lines forming an X pattern. As the shots move position, different sections of the geophone lines are made live by the recording instruments in order to maintain the same relationships between sources and receivers as in fig. 2.17. Figure 2.18 shows part of a full line swathe with a full shot pattern in place. The whole arrangement is then moved for the next set of six receiver lines (fig. 2.19). In this case there is an overlap of one line between one set of six receiver lines and the next set. Special software is available to the designer of land seismic surveys to allow determination of the fold, azimuthal coverage and distribution of source–receiver distance at any location. Balances need to be struck between the amount of equipment available and the need to continually pick up and move cables and receivers. Usually in land acquisition it is not economic to get the simple uniform geometry of marine operations and often there is a small overprint of the acquisition design in the final processed data.

In areas that are difficult for vehicle access (e.g. mountains, forests), an explosive source is used. Usually the area covered per day is not as great as achieved by vibrator trucks. Holes need to be drilled into the earth to ensure the explosive charges are well coupled to bedrock and not fired in the shallower weathered layers, which would cause excessive amounts of noise in the data. This reduces the acquisition rates to somewhere between 50 and 100 shotpoints per day. The lack of vehicles may mean the majority of movement of equipment has to be by helicopter, an important issue for safety. Another issue is the amount of time that may have to be spent cutting lines in forested areas.

Whatever type of land source is used, the entire survey needs to be corrected for the arrival time changes due to both topography and variations in the thickness of the near-surface layer (fig. 2.20). The near surface is generally heavily weathered and usually has altered acoustic properties compared with the less affected deeper layers. This generally results in a much slower velocity, but in areas of permafrost the near surface may be substantially faster. Usually a separate crew is responsible for drilling and measuring uphole times (the time required to travel from the bottom of the drilled hole to the surface) throughout the area. These are then used to determine the depth to the base of the weathered layer, and the velocities in the weathered and bedrock layers from which the static corrections can be derived. Sheriff & Geldart (1995) give an introduction to these corrections. It is impossible and uneconomic to drill holes at every shot and geophone location, so first arrival times from the field records are also used to estimate static corrections. In some circumstances, special refraction profile crews are used to determine the near-surface velocity profile. When calculating the statics to be applied, it is important to ensure consistency across the area. The large redundancy in 3-D data means that there is often conflicting information about the statics required at any one location and special software is used to generate the corrections.

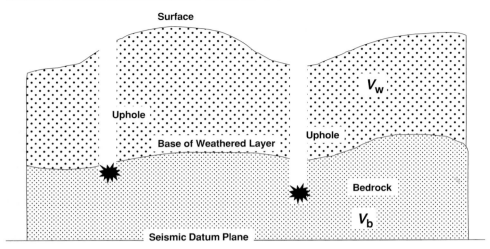

Fig. 2.20 Weathering layer corrections for land data. For land acquisition the surface topography and thickness of the weathered zone must be corrected for, since otherwise the underlying structure would have an overprint of elevation and weathering layer changes. Uphole information together with refraction records is used to build a model of the near surface. Once the velocity of the bedrock and weathering layer and the depth to the base of the weathering layer have been determined, the correction is simply the sum of the shot and receiver statics. The static at any location is simply $(E_s - E_w)/V_w + (E_w - E_d)/V_b$, where E_s is the surface elevation, E_w is the elevation of the base of the weathered layer, E_d is the elevation of the seismic datum, and V_w and V_b are the velocities in the weathered and bedrock layers.

2.4   Other types of seismic survey

Marine surveys require a minimum depth of water and land surveys must terminate in areas close to the shore. The zone in between is termed the transition zone and surveys involve a variety of specialised equipment such as specially designed swamp buggies with shallow draughts. Although the operation in these areas is more time consuming and has some extra constraints such as the timing of tides, they can be considered as a mixture of marine and land operations.

4-D seismic surveys are repeat seismic surveys, generally over a producing field, that are specially shot at incremental time periods to determine differences in seismic response due to production. The fourth dimension referred to in the title is time, the other three dimensions being the standard ones of two horizontal co-ordinates and one vertical spatial co-ordinate. Both pressure and fluid within the reservoir may change during production and both of these affect the acoustic properties of the rock. Even the absence of a change in part of the reservoir may indicate compartmentalisation due to faulting, etc. These 4-D surveys may be land or marine, with one to four component receivers, although the majority to date have been conventional towed-streamer marine surveys. Jack (1997) gives an excellent introduction to the subject, which is considered in more detail in chapter 8.

2.5   3-D data processing

As with the section on data acquisition, we do not aim to give here a complete overview of seismic data processing. Rather, we will again concentrate on the differences between 2-D and 3-D. Yilmaz (1987) gives a detailed but very readable account covering all aspects of data processing.

In most respects 3-D data processing is very similar to 2-D data processing. Indeed, any process that treats traces individually, such as statistical deconvolution, gain recovery, trace muting and frequency filtering, is applied in an identical manner in 3-D and in 2-D. Some of the other processes are applied in a 2-D sense, along lines rather than across the 3-D data volume.

There are three main processes that are different for 3-D data: binning, spatial filtering, and any process that repositions data such as migration and dip moveout. In addition several other applications, although generally run on selected lines from the 3-D volume, may benefit from the prior application of one of the above. An example of such a process is velocity analysis, which is normally run on a regular grid of 2-D lines but will improve in quality after 3-D migration. An important aspect of 3-D data processing is to ensure spatial consistency of parameters such as the velocity field.

Figure 2.21 shows a typical processing sequence for 3-D data. We will briefly run through all the steps, but those that are significantly different from 2-D processing are highlighted in bold and will be discussed in more detail.

Modern commercial processing systems contain hundreds of algorithms. There are often several ways to tackle any given problem and the processing analyst may have to try several different techniques and parameters before being satisfied that any particular step is optimised. It is not possible in this book to even introduce all of the techniques but the reader is referred to Yilmaz (1987) and Hatton *et al.* (1986) for more detailed descriptions.

2.5.1   Reformat, designature, resampling and gain adjustment

The first step in fig. 2.21 is a reformatting operation. This just takes the data coming from the receivers and puts them into trace order. Normally the data are written to tape at this stage in one of the designated industry formats so that the raw records are retained to form the basis of possible later reprocessing. If starting from field tapes, reformatting includes converting the data from standard industry format into whatever format the processing system uses. The second step, designature, takes the wavelet

1. Reformat
2. Designature
3. Resample from 2 to 4 ms
4. Low cut filter (5/12 minimum phase filter)
5. Remove bad traces
6. Merge navigation with seismic headers
7. Spatial resampling from 12.5 to 25 m groups. Normal moveout (NMO) correction, K-filter, trace drop, inverse NMO
8. Spherical divergence gain corrections
9. Deconvolution before stack
10. **Shot interpolation to double fold in CMP gathers**
11. Radon demultiple on interpolated gathers
12. High-frequency noise removal
13. Drop of interpolated traces
14. **Flex binning increasing from 37.5 m on near to 50 m on far offsets**
15. Sort to common offset
16. Dip moveout (including approximate NMO) halving the number of offset planes on output
17. **Pre-stack time migration using constant velocity 1600 m/s**
18. Inverse NMO
19. Re-pick velocities (0.5 km grid)
20. NMO

Processing hereafter continuing on three volumes:
21. Stack to generate three volumes: near offset stack, far offset stack and full offset stack
22. 3-D constant velocity inverse time migration
23. Bulk static (gun and cable correction)
24. K-notch filter to remove pattern caused by the acquisition
25. **FXY deconvolution**
26. **FXY interpolation to 12.5 m × 12.5 m bin grid**
27. Pre-migration data conditioning (e.g. amplitude balance, edge tapers, etc.)
28. **One pass steep dip 3-D time migration (using time and spatially varying velocity field)**
29. Zero-phase conversion by matching to wells
30. Spectral equalisation
31. Bandpass filter
32. Residual scaling

Fig. 2.21   Typical 3-D processing sequence.

that was created by the source and converts it to a more compact form. For instance, air guns output a signal with a main peak, followed by a smaller secondary peak due to the re-expansion of the air bubble. Such a source signal is undesirable since every reflection would be followed by a smaller repetition of itself. Designature removes the second peak, giving the input wavelet a more compact form. A decision needs to be made at this point whether the output should be zero- or minimum-phase. A zero-phase wavelet is one that is symmetrical about its centre, while a minimum-phase wavelet is one that starts at time zero and has as much energy near the start as physically possible (fig. 2.22). There are arguments on both sides. It is certainly desirable to use zero-phase

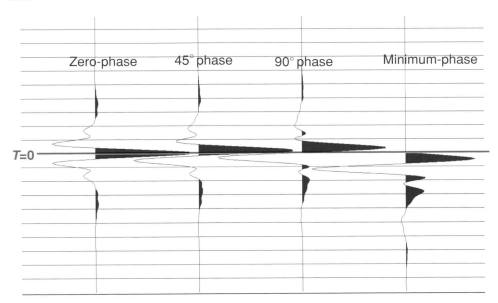

Fig. 2.22   Comparison of zero-phase and minimum-phase wavelets. Also shown are zero-phase wavelets that have had their phases rotated by 45° and 90°; the latter develops an antisymmetric form. The desired output from seismic data processing is usually a seismic section that represents the earth reflectivity convolved with a zero-phase wavelet, because such a wavelet has the greatest resolution for any given bandwidth. Seismic sources cannot be zero-phase since that would imply output before time zero. Most air guns give signatures that are close to minimum-phase. This has the maximum amount of front-loading of the wavelet possible for any given amplitude spectrum. During processing the data are converted to zero-phase.

data since later processes such as velocity analysis will benefit from such data. However, the earth attenuates higher frequencies in the wavelet as it passes through it, and this alters the phase of the wavelet. Even if the wavelet going into the ground were to be zero-phase, it would not remain so at depth. To combat this, if the zero-phase option is chosen, then one needs to account for the phase distortion with depth by also applying a deabsorption filter. If the minimum-phase option is chosen then the application of the deabsorption filter is usually neglected, since absorption itself is a minimum-phase process. Thus, the wavelet remains minimum-phase (but since it is not a constant phase wavelet, its change with depth makes it impossible to zero-phase the entire section later on). Figure 2.23 shows a wavelet before and after designature.

Marine data are normally recorded with a 2 ms sampling interval. This is sufficient to record frequencies up to 250 Hz, much higher than the frequencies that are actually recorded from the earth, particularly at the deeper target levels where it is uncommon to record frequencies higher than 30–50 Hz. The data are usually resampled to 4 ms, which is sufficient for frequencies up to 125 Hz. An anti-alias filter is applied to ensure that any higher frequencies in the near surface do not alias on to lower frequencies. This resampling reduces the volume of data by 50% and speeds up all of the later processing stages.

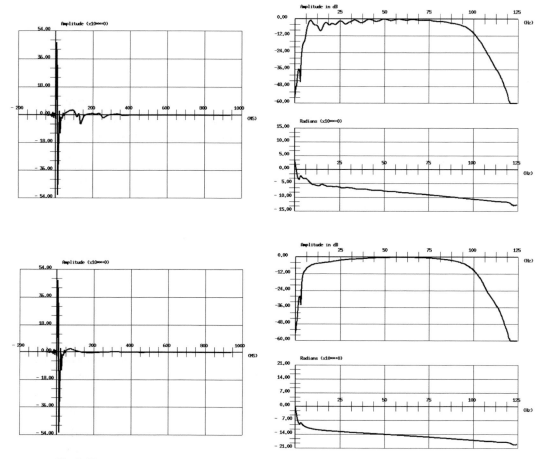

Fig. 2.23 Example of a wavelet before (top) and after (bottom) designature. The source from the air gun array still contains remnant bubble oscillations. A filter is applied to remove these periodic events from the recorded seismic. The sharpening of the wavelet at this stage gives increased resolution in the seismic data. In this case, no attempt has been made to convert the wavelet to zero-phase.

Step 4 removes the lowest frequencies from the data. These are often heavily contaminated by noise (e.g. swell noise in the marine case), which may swamp any underlying signal and is best removed. Step 5 is an editing step; any traces that appear excessively noisy owing to poor coupling or equipment failure are removed. Step 6 is the important process of assigning the correct positioning information to the traces as referred to in the acquisition section, so that actual offsets and locations can be determined.

Just as step 3 was associated with resampling the data in time, then step 7 is an identical process in space. Typically the receiver group interval within a marine streamer is 12.5 m. This gives a CMP interval of 6.25 m, which in all but the steepest dip areas gives an over-sampling in the inline direction. The data are merely resampled to mimic

25 m receiver group intervals, with every other trace being discarded. As with the resampling in time, an anti-alias filter is applied before this step. The resampling further reduces data volumes, by a second factor of two.

Step 8 applies a time-varying gain to the data to boost up the amplitudes of the later arrivals compared to the earlier ones. As the wavefront from the source travels deeper into the earth, it covers a larger area and also suffers amplitude decay due to transmission losses and attenuation. Spherical divergence correction is applied to remove the loss in amplitude due to the wavefront expanding with depth. This expansion means that the same energy in the wavefront is spread over an increasing area as the distance travelled by the wave increases, and hence the amplitude of the wave is less. One can see the same effect (in a 2-D sense) on waves caused by dropping pebbles into water. Near where the stone was dropped in, the perimeter of the circular wavefront is quite small and its amplitude is large, but as the wavefront expands its amplitude decreases. In addition to a spherical divergence correction it is common to apply an additional exponential gain function with time to account for the transmission and attenuation losses.

2.5.2   Deconvolution

Step 9, deconvolution, is a process that sharpens the wavelet and removes any short period reverberations. The theory is explained in a number of texts; see, for example, Robinson and Treitel (1980). A digital operator is designed for each trace that is convolved with the trace to remove unwanted ringiness. The operator is designed automatically based on characteristics of the traces and a few simple parameters supplied by the processing analyst. These include the operator gap. The idea is that the operator will not change the wavelet from time zero to the end-time of the gap, but try to remove periodicity at times beyond the end of the gap. Deconvolution may be predictive or spiking, with the difference between the two being the length of the operator gap. A short, or no gap, gives maximal wavelet compression (hence the name spiking) while a large gap (32 ms or more) attempts to remove periodicity caused by short period multiples whose period is longer than the gap. In practice, both forms actually perform a mixture of wavelet compression and de-reverberation. Care should be taken with spiking deconvolution since theoretically it assumes minimum-phase input data. The output of data that have undergone too severe a spiking deconvolution process is relatively variable in phase. Because of this, it is becoming rare to apply spiking deconvolution before stack. Consistency of wavelet shape and amplitude becomes increasingly important as more attempts are made to infer subsurface information from the amplitudes of the reflection events.

2.5.3   Removing multiples

One of the consequences of dual source shooting is that the number of traces in a common midpoint (CMP) gather is half that of single source shooting. Each source

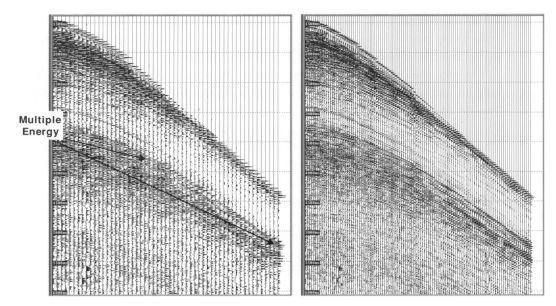

Multiple Energy

Fig. 2.24 CMP gather at 100 and 50 m trace spacing. Comparison of CMP gather at original trace spacing and after shot record interpolation (see fig. 2.25) to interpolate traces so that the new gather has half the trace spacing of the original. The high-energy multiple packet is badly aliased in the original gather, making removal of these multiples very difficult. After shot record interpolation the multiples are better sampled and there is little aliased energy.

contributes to data on different subsurface lines (fig. 2.6). The time interval between successive shots is unchanged from the single source case, because a source cannot be fired until the listening period for the previous shot has ended. When the dual sources are fired alternately, there are therefore half as many shots contributing to every CMP. This may lead to data aliasing of any steeply dipping multiples within a CMP gather. For example, a common geometry is to have 25 m receiver group intervals within a cable and two sources fired alternately, with a shot every 25 m. This means that each source is fired only every 50 m. The fold in a CMP gather will be only one-quarter of the number of traces in a shot gather, and the offset increment in a CMP gather will be four times that in a shot gather. Figure 2.24 shows an example CMP gather at 50 and 100 m trace spacing. To reduce the possibility of aliasing, additional shots are created by interpolation between the actual shots. Generally, frequency-space methods are employed since these use the lower unaliased frequencies to generate estimates of the higher frequencies. The procedure is explained by the stacking diagram in fig. 2.25. The shot recorded data are sorted into receiver gathers; each receiver gather is then interpolated to add one new trace between each existing trace. These traces are then used together with the original data to form better-sampled CMPs. Usually shot interpolation is required for multiple suppression and when this has been performed the additional (invented) traces are discarded.

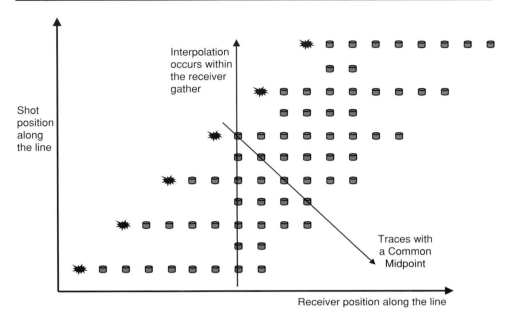

Fig. 2.25 Stacking diagram to illustrate shot record interpolation. The grey receivers mark the position of the original recorded data. The data are sorted into receiver gathers in which the trace spacing is half what it is in CMP gathers. Interpolation is used to create the additional traces shown in red. Once sorted into CMP gathers the number of traces within each gather has doubled and the trace separation has halved.

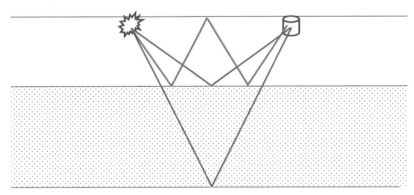

Fig. 2.26 Primary and multiple reflections. The red and purple events are primaries; they have a single reflection along the ray-path. The green event has multiple reflections and in this case is the first-order multiple of the purple event. The timing of the green event may be similar to underlying primaries, and if it is not removed it may obscure the deeper reflectivity.

The next step in fig. 2.21 is the process to remove multiples from the data. Figure 2.26 explains the differences between primary events and multiples. Here again there are a large number of choices of technique. Methods are based either on velocity move-out differences, or on prediction based on the timing and geometry of the cause of the multiple. In the former class are the most commonly used techniques, Radon and FK demultiple. In both cases the data are normal moveout (NMO) corrected to approximately flatten either the multiple or primary events. (The concept of NMO is explained in Appendix 2 and the equations for it are given in section 3.3.3.) The data are then transformed to Radon or FK space and the unwanted events are removed by rejecting (or passing and subtracting) the multiple events. The methods work because multiples spend longer travelling in the (generally slower) near-surface and hence have a moveout velocity that is slower than the primary events arriving at the same time (fig. 2.27). Multiple elimination in areas of strong water bottom reflection

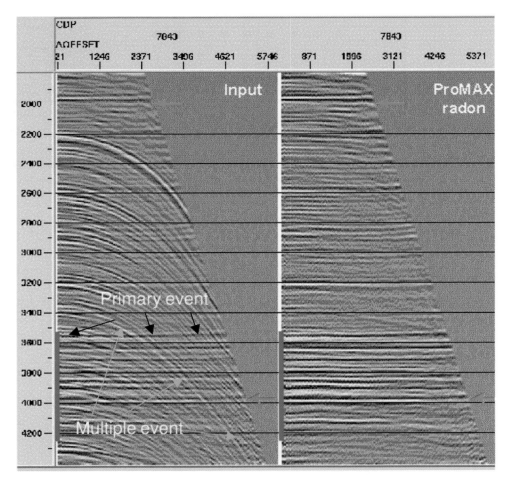

Fig. 2.27 Gather showing primary and multiple events, illustrating moveout differences between the two. In this example the primary NMO velocity has been used to flatten the primary events. The multiples, in this case from a deep seabed, are dipping from top left to bottom right. This example shows the removal of the multiple by the Radon transform technique, where events with moveout on the farthest offset greater than 100 ms and less than 1500 ms were removed.

coefficients or with large velocity contrasts (such as sub-salt plays in the Gulf of Mexico) is an area of active development within the geophysical industry. Recently introduced wave equation methods are based on the prediction of the multiple from the primary location. These techniques show considerable promise but they are expensive compared with moveout techniques and do not easily lend themselves to true 3-D implementation owing to acquisition limitations. Currently they are applied to data in a 2-D sense and fail if there is a strong cross-dip component in the multiple-generating surface.

2.5.4   Binning

Data binning is the process where every trace is assigned to midpoint locations (i.e. a location halfway between shot and receiver). To achieve this the survey area is covered by a large number of rectangular bins as shown in fig. 2.28. Traces are then assigned a Common Midpoint (CMP) location at the centre of the bin in which they fall. In marine acquisition, if the streamers were perfectly straight behind the boat, all the bins would contain a regular sampling of offsets, which would be highly desirable for seismic processing. However, because of feathering this is not the case. In particular, the

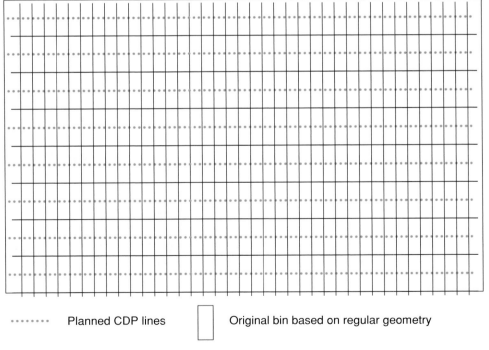

· · · · · · · · Planned CDP lines   ☐ Original bin based on regular geometry

Fig. 2.28   Data binning. The size of the bins is determined by the original acquisition. The bins are centred on the planned locations with width equal to the CMP (also known as CDP) spacing and length equal to the line spacing.

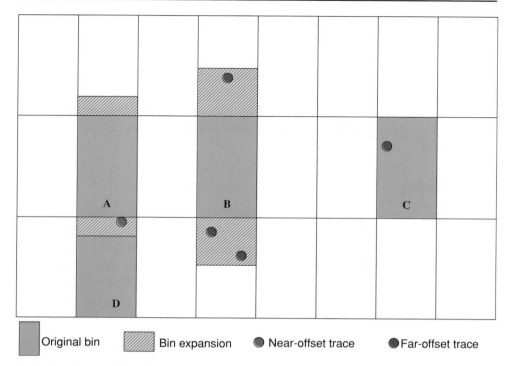

Original bin   Bin expansion   ● Near-offset trace   ●Far-offset trace

Fig. 2.29 Bin expansion. Flexi-binning is used so as to ensure that each bin has a regular distribution of traces with the full range of offsets. Three examples are illustrated here. Bin A is expanded for a near offset trace with relatively small expansion. Bin B is seeking a trace with larger offset. The bin expansion is greater and has thrown up three possibilities. The trace closest to the bin centre is the one that gets borrowed. Bin C already has a far offset trace and no bin expansion is required. The red trace will end up being used by both bin A and bin D. After binning, the traces are treated as though they came from the centre of the bin though it is normal to retain and use the absolute offset information for later processing.

furthest offsets may be non-uniformly distributed with some bins having two or more traces with a similar offset while others have none. To make the distribution of offsets more uniform a process called flexi-binning is used. Here the bins are extended by fixed amounts (generally) in the crossline direction, perhaps by as much as 50% either side (fig. 2.29). If no trace within a certain offset range is found within the original bin, then a trace within the extended bin is used as though it falls in the original bin. If more than one trace within an offset range is found, then it is usual for the trace closest to the bin centre to be used. The bin expansion is normally small for the near offsets, since the cable position is better controlled close to the boat, and grows with increasing offset. Note that the process as described just copies a trace from one location to another. This copying means that there may be small static shifts between traces within a binned CDP, since one trace may come from the extreme top edge of a flexed bin and the next offset may come from the bottom edge. Figure 2.30 shows a binned gather with

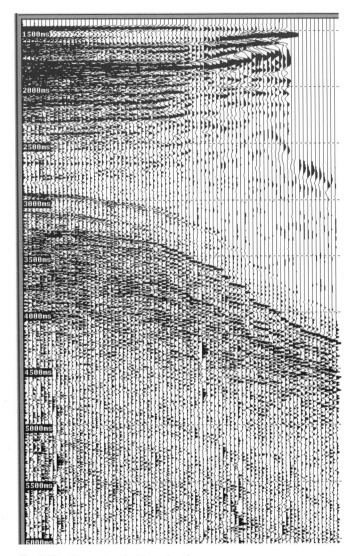

Fig. 2.30 Example of a binned gather.

some clear differences between adjacent traces. This may substantially reduce the final frequency content of the stacked data. Some contractors offer full interpolation of traces onto bin centres or full pre-stack migration procedures that can correctly handle the real trace positioning rather than assuming traces are regularly sampled at bin centres, both of which provide for a mathematically better solution. Owing to the cost this is not routinely applied at present, but with the ever-reducing cost of computer power and the desire for higher-frequency data for improved interpretation it is becoming more common.

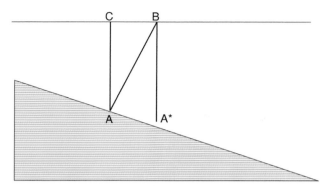

Fig. 2.31 Seismic migration. During seismic data recording the reflection from point A will be recorded by a coincident source and receiver at location B. It will be plotted on a stacked section as a point A* vertically below B with distance given by length AB. Migration moves the point back to its subsurface reflection point and plots it vertically below point C, thus giving a section that looks more like a cross-section through the earth.

2.5.5   Stacking and migration

The next steps in the processing sequence are to prepare the data for seismic migration. Migration is one of the key steps in seismic data processing – it is the step which attempts to move the recorded data so that events lie in their correct spatial location rather than their recorded location (figs. 2.31 and 2.32). The mathematics of how to perform this process is well defined by the wave equation and has been extensively researched over the past three decades. There are a number of texts, such as Stolt & Benson (1986), Berkhout (1982, 1984) and Bancroft (1997, 1998), specifically written about seismic migration that describe the mathematics in detail. Once more there are a large number of options ranging from migrating all the pre-stack data to stacking data in a CMP followed by post-stack migration. There is also the issue of whether to use time or depth migration and also the type of algorithm (Kirchhoff, implicit finite difference, explicit finite difference, FK, phase shift, etc.). In recent years the choice has become even wider with the ability of some algorithms to incorporate the effects of velocity anisotropy. To a large extent the details of the algorithm are unimportant to the interpreter. What matters are the accuracy and cost. These are determined by issues such as the largest dip that can be properly migrated, the frequency content of the final migration and the time required to perform the operation. The choice of whether to migrate data before or after stacking is largely dependent on the velocity regime and the subsurface dips present in the data. Large dips may mean that shallower (slower) events arrive at the same time as deeper events, giving rise to the two events needing to stack at the same time but with different velocities (fig. 2.33). Such an occurrence is known as a stacking conflict and the solution is to migrate the data before stack. The

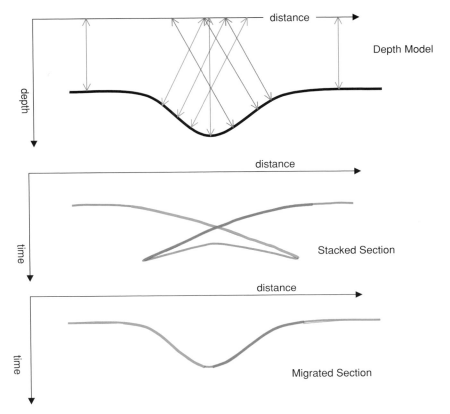

distance

Depth Model

depth

distance

time

Stacked Section

distance

time

Migrated Section

Fig. 2.32   Seismic migration moves events from their recorded position to their true subsurface position. This diagram shows the effect of migration on a syncline. The rays are reflected at right angles to the surface. It can be seen that the blue rays from the right hand of the syncline are sometimes recorded to the left of the green rays reflected from the left side of the syncline. This gives a complicated bow tie pattern in the corresponding time section. The effect of migration is to move the events back to their true subsurface location.

earliest forms of migration were based on an approximation to the wave equation and ignored ray-bending at velocity boundaries. Today such techniques are known as time migration methods. In areas of lateral velocity contrasts the straight ray approximation of time migration can seriously misposition events. The solution to this problem is depth migration, which correctly handles velocity variation. Of course, there is the additional complication of deriving the velocity model needed to drive depth imaging and this is another active area of research and development. Most methods require an iterative approach. An assumed velocity–depth model is used, the data are migrated pre-stack using this model, and the images across the migrated CMP gather are compared. If the model velocity is too high then the further offsets will have the event positioned too deep compared with the near offsets, and conversely if the velocity is too low the events will be imaged shallower on the far offsets than on the nears. Only if the velocity

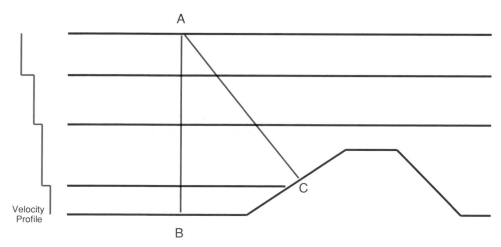

Fig. 2.33 Example of events causing a stacking conflict. Reflections from both B and C will be recorded at the point A with approximately the same travel time. Since event C is higher in the section it requires a slower velocity to stack the event than the deeper reflection B. This creates a stacking velocity conflict that is solved by partially migrating event C away from event B. Provided the migration velocity is slower than the true earth velocity the partial migration will move the dipping event to somewhere between its unmigrated location (B) and its true location (C).

model is nearly correct will the events appear flat. Figure 2.34 shows an iteration from a pre-stack depth migration where velocity analysis is being performed. Events will only be flat if they have been fully migrated in a 3-D sense and it is expensive to repeat this for the entire section. Usually the velocity iteration step is performed only on a series of target lines using an algorithm (such as one of the variants of Kirchhoff) that can output onto a discrete grid without having to migrate the entire dataset. Since lateral velocity variations also give rise to stacking problems most depth migration benefit is gained from working pre-stack.

In the early days of 3-D data, migration was performed as two runs of 2-D migration. First the inlines were 2-D migrated, and then the data would be sorted into crosslines and migrated again. This was known as two-pass migration and is relatively common on older surveys. It was used because computer power and memory were not large enough to allow the full 3-D dataset to be migrated in one pass. The two-pass technique is theoretically correct only for a constant velocity earth, and once computers could hold sufficient data to allow full 3-D migration it actually became cheaper to perform the migration in one pass rather than the two because it avoided the sort from inline to crossline orientation. Today it is extremely rare for 3-D migration to be performed as two 2-D migrations.

The migration approach outlined in fig. 2.21 is a hybrid pre-stack time migration approach and covers steps 15 to 28. The processing sequence comes from an area with relatively steep dips but without large lateral velocity variations. Therefore, pre-stack

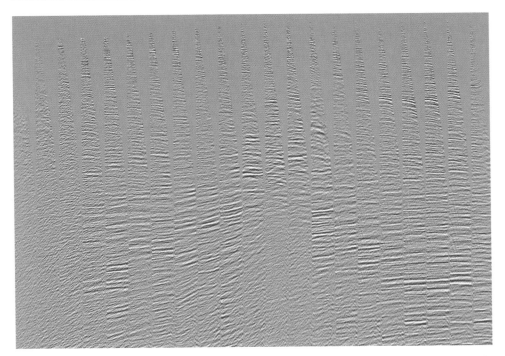

Fig. 2.34 Examination of CMP gathers during an iteration of pre-stack depth migration. Pre-stack depth migration is required in areas of complex lateral velocity variation. In such areas the standard normal moveout equation is insufficient to align events owing to the different paths that the waves take through the subsurface. In order to apply pre-stack depth migration one needs to supply the correct velocity–depth model. This is generally not known and so is built in an iterative fashion during the pre-stack migration. The correct velocity model is the one where the migrated gathers are flat across all offsets. Here we see gathers during a pre-stack migration iteration. Some of the events are flat, indicating that the correct model has been determined. Others are still dipping and the velocity model needs updating to flatten them also.

time migration is the preferred solution. Since full pre-stack time migration is still relatively expensive a cheaper, faster approximation has been used. As computer power continues to increase, these approximations are being replaced by the full pre-stack migration solution. Steps 15 and 16 are an application of Dip Moveout (DMO), so called because it removes the effect of dip on stacking velocities and trace positions. As part of the DMO process, one also needs to apply an approximate NMO correction. At this stage it is usual to have picked velocities approximately on a sparse grid, say every 1 km in all directions. This is generally sufficient since detailed velocity analysis for final stacking (maximum analysis separation of 0.5 km) is performed after DMO and constant velocity migration in step 19. There is an excellent article on the benefits of DMO by Deregowski (1986) and more detail on algorithms in the notes by Hale (1991). One of the benefits of DMO is that it takes a constant offset section and transforms the data to a zero-offset section, thus allowing conventional post-stack migration to be applied. This is how the pre-stack migration process in step 17 is applied. In this step, we are not attempting to get a final migration; instead we are trying to move events with potential stacking conflicts apart sufficiently so that we can determine a unique stacking velocity. This allows us to use a computationally very fast constant velocity form of migration – in this example performed with a velocity of 1600 m/s. It is important that this step does not use too high a velocity since then events can disappear completely if they are steeply dipping. Generally one uses a velocity that is close to the slowest seen in the section (1480 m/s for water velocity). Step 18 removes the approximate correction for NMO applied during the DMO process (step 16) and this allows for the detailed velocity analysis to be performed and applied in steps 19 and 20. The data are usually saved to tape after DMO and the removal of the initial NMO correction to allow any later re-processing to begin from this stage. It is common practice to perform the final velocity analysis at a spatial interval of 250–500 m though this too is reducing in the desire to retain the higher frequencies. The spacing needed depends on the variability in the velocity field; large variations require more frequent analysis than more gradual changes. For really detailed work, the advent of better automatic velocity picking algorithms means that it is possible to pick velocities for every CMP location within the target area.

After the final velocity analysis and moveout correction the data are stacked. Stacking together traces that contain the same reflection information both improves the signal to (random) noise content (by the square root of the number of traces stacked) and reduces any residual coherent noise such as multiples which stack at velocities different from the primary events. During stacking, mutes (zeroing the data within specified zones) are applied to the data to ensure that NMO stretch is not a problem and that any residual multiples left on the near-offset traces do not contaminate the stacked section (fig. 2.35). There may be some amplitude variation with offset (AVO) effects in the data, which can be used as hydrocarbon indicators, so this information is retained by stacking the data over selected offset ranges. Chapter 5 contains more details on the physics and use of AVO data. An offset to angle relationship is used that depends on the subsurface velocities to define mute patterns to produce two or more angle stacks as well as the stack over all usable offsets. All processing from this point onwards needs to be applied to each of the individual stacked sections.

The result of the stacking after step 17 is a number of partially migrated sections requiring the removal of the constant velocity migration. This is done by step 22, an inverse migration or demigration using the same constant velocity as used for the initial migration.

Step 23 adds a static correction based on the depth of the source and the receiver so that the datum is now at sea-surface. Step 24 removes some of the effects of the pattern left in the data from the acquisition variability. Since different CMPs will contain a different combination of traces in a regular pattern this may show itself in the final

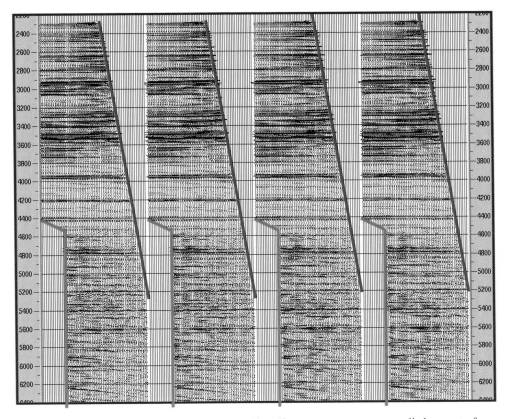

Fig. 2.35   Example of mutes applied during stacking. Here we see two mutes applied to a set of gathers. The red mute is applied to stop events being distorted by the stretch caused by the normal moveout process at large offsets and shallow times. The green inner trace mute has been applied to remove remnant multiples on the near traces that have not been removed by other demultiple processes such as Radon.

stacked section. For instance, CMP 1 may contain traces 1,5,9..., CMP 2 will have offsets 2,6,10..., CMP 3 offsets 3,7,11..., CMP 4 4,8,12..., CMP 5 will be back to 1,5,9.... This four-trace pattern may be visible in the final section particularly in the near surface where the fold of data (number of traces stacked together) is low owing to the stretch mute. Any regularly repeating pattern will show itself as a strong component in the inline spatial Fourier transform and can be removed by a notch filter.

Step 25 is a 3-D noise removal process. The application of noise suppression is much more successful with 3-D data than 2-D owing to the extra dimension, data volume and spatial consistency for the algorithm to work with. Usually relatively small operators are used, working on maybe five traces in both directions. This allows the filters to adapt to relatively sudden changes in reflector dip while retaining sufficient data to distinguish signal from random noise. Step 26 is very similar to step 10, the difference being that we are now performing the operation post stack in order to make the crossline spacing the same as the inline spacing. As already discussed, the CMP spacing along a line is generally much finer than the spacing between lines owing to cost and the possibility of cables becoming entangled. Again methods that attempt to interpolate frequencies beyond the standard aliasing criterion are employed since no benefit would be gained otherwise. Both steps 25 and 26 together with step 27 are used to condition the data for the final migration process of step 28. Within step 27 we are attempting to remove any sudden amplitude variations between traces since these are not consistent with the behaviour of waves and would cause artefacts during the migration process.

Step 28 is the final, high-fidelity migration process using the spatial varying velocity field from step 20. It is usual not to use the raw velocities as originally picked, but rather to use velocities that have been smoothed both spatially and in time. Often trials with various percentages of the picked velocity field are performed and the results are analysed for evidence of over (too high velocities) or under (too low velocities) migration. A composite velocity field is used for the final migration based on these tests. The result of 3-D migration is often very impressive compared to 2-D migration. The reason is simple: 3-D migration places the energy correctly in a three-dimensional sense, whereas 2-D migration can only do this in the direction the data were shot; any complexity perpendicular to the plane of the 2-D section remains unresolved. Figure 2.36 shows an example of seismic data before and after migration. The effect of the migration is to narrow the width of the large anticlinal feature and steepen the dips of the sides.

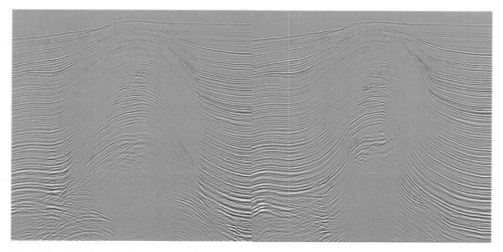

Fig. 2.36 Example of seismic data before and after migration. Data on the left are unmigrated, on the right they are migrated. Notice how the migration process has reduced the apparent size of the structure. Note also that whereas the picture on the left has reflections that cross each other, those on the right are geologically consistent.

2.5.6 Post-migration processing

After migration the data are more correctly positioned although they may still not be perfectly located owing to errors in the velocity field, use of time migration rather than depth migration, anisotropy, etc. However, at this stage the data should look like a cross-section through the earth with reflections corresponding to changes in acoustic properties and unwanted events such as multiples and noise removed. The final step is to convert the data to zero-phase which centres the peak of the seismic wavelet on the impedance contrast. Even when an attempt has been made to keep the data zero-phase throughout the processing sequence, it is likely that there is still considerable phase uncertainty due to the effects of attenuation. Zero-phasing is best performed with well data. An operator is determined by least-squares matching the seismic data to the reflectivity generated from the well data (White (1980) and Walden & White (1984)). Since there is likely to be some error in the positioning of the seismic data after the migration, this matching is usually done for a number of traces surrounding the well location. The fit between the seismic trace and the well log generated synthetic is compared and the trace location that gives the best match is chosen as the trace for estimation of the wavelet. The process is explained in more detail in chapter 3 (section 3.1.1). Figure 2.37 shows the wavelet extraction obtained from one of the commercially available interpretation packages. Once a wavelet has been extracted an operator can be designed to convert it, and hence the data, to zero-phase. Note that such an operator is really only applicable to the time-gate used for the extraction. It is extremely difficult to confidently zero-phase data over large time windows.

The final three steps in the processing sequence outlined in fig. 2.21 are all concerned with fine tuning the data prior to loading to the interpretation workstation for detailed analysis. Step 30 applies an equalisation on a trace by trace basis to ensure the spectral content of each trace is broadly similar, while step 31 applies time-varying bandpass filters to reduce the higher frequencies with time to eliminate those that are mostly noise due to their attenuation on passing through the rocks. Finally, step 32 applies time-varying trace scaling to ensure a balanced-looking section with time. One approach is to apply Automatic Gain Control (AGC). This applies a time-varying gain to each trace individually, with the gain calculated so as to keep the average absolute amplitude constant within a window that slides down the trace. A short time-window AGC (say 200 ms) is highly undesirable if any use will be made of amplitude information subsequently, because it tends to destroy lateral amplitude changes that may be important. A long gate AGC (say 1000 ms or more) is usually acceptable, however, because the gain is not much influenced by the amplitude of any single reflector. All these final steps are important, because they can aid or inhibit the interpretability of the dataset that is delivered to the interpretation workstation. Remove too many high-frequency data and subtle detail may be missing; leave too much noise and automatic batch tracking of horizons may be compromised.

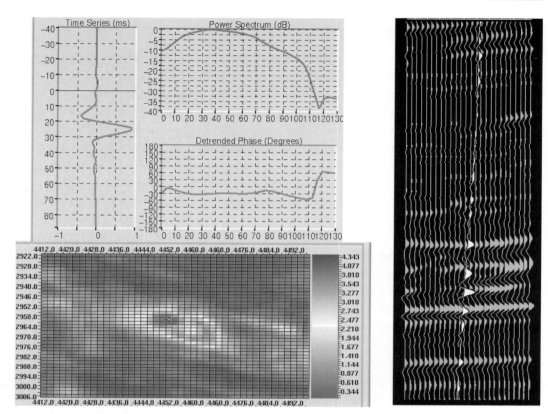

Fig. 2.37   Example of wavelet estimation by matching data to wells. The wavelet may be estimated by deriving a matching operator between the seismic data and the well log predicted reflectivity. To do this a small area of traces centred on the well location is examined and the match between each trace and the well log is calculated. The best match trace is determined and the wavelet is estimated together with the synthetic that is predicted using the wavelet and log derived reflectivity. Several QC parameters are also given, the most important being the signal to noise estimate. The bottom left-hand display shows the signal to noise plot for traces around a well; high signal to noise is shown in blue. In this case we have a very good match with a signal to noise ratio of over 4 that is obtained from a trace very close to the actual well location. The display on the right shows the synthetic in white overlain on the recorded seismic data.

With luck, hard work and appropriate testing and parameter selection the result of the complete processing sequence should be a section that resembles a cross-section of the reflectivity through the earth with optimised seismic resolution and signal to noise ratio. In practice it often takes several passes through the processing sequence before this goal is satisfactorily achieved. Each pass through the data uses lessons learnt from the previous processing route so that the next iteration may spend more time on velocity analysis or multiple attenuation or use a different migration strategy. There is also a different requirement from the data as we move from exploration through appraisal and development. Once exploration targets have been identified there may be a wish for a detailed geophysical analysis of the prospective levels. Techniques such as AVO analysis, seismic inversion and amplitude analysis (see later chapters) all impose special requirements on the quality of the seismic data that may not be met by the initial processing for exploration screening.

In some areas with severe problems (e.g. sub-basalt or beneath complex salt bodies), where the primary signal is strongly attenuated or scattered, it may be difficult or impossible to generate interpretable sections with current technology. The desire for better seismic sections remains strong and there is active research in most aspects of seismic data acquisition and processing. However, the industry should be proud of the improvements made in seismic technology over the past two decades. Data quality has improved tremendously and with it our ability to see hydrocarbons at depth.

References

Bancroft, J. C. (1997). *A Practical Understanding of Pre- and Poststack Migrations, Volume 1 (Poststack)*. Society of Exploration Geophysicists, Course Notes Series No. 7.

Bancroft, J. C. (1998). *A Practical Understanding of Pre- and Poststack Migrations, Volume 2 (Prestack)*. Society of Exploration Geophysicists, Course Notes Series No. 9.

Berkhout, A. J. (1982). *Seismic Migration. Imaging of Acoustic Energy by Wave Field Extrapolation. A. Theoretical Aspects* (2nd edn). Elsevier Scientific Publishing, Amsterdam.

Berkhout, A. J. (1984). *Seismic Migration. Imaging of Acoustic Energy by Wave Field Extrapolation. B. Practical Aspects*. Elsevier Scientific Publishing, Amsterdam.

Deregowski, S. M. (1986). What is DMO? *First Break*, **4** (July), pp. 7–24.

Evans, B. J. (1997). *A Handbook For Seismic Data Acquisition In Exploration*. Society of Exploration Geophysicists, Geophysical Monograph Series No. 7.

Garotta, R. (1999). *Shear Waves from Acquisition to Interpretation*. Society of Exploration Geophysicists, Distinguished Instructor Series, No. 3.

Hale, D. (1991). *Dip Moveout Processing*. Society of Exploration Geophysicists, Course Notes Series No. 4.

Hatton, L., Worthington, M. H. and Makin, J. (1986). *Seismic Data Processing: Theory and Practice*. Blackwell Scientific, London.

Jack, I. (1997). *Time-Lapse Seismic in Reservoir Management*. Society of Exploration Geophysicists, 1998 Distinguished Instructor Short Course.

McQuillin, R., Bacon, M. & Barclay, W. (1984). *An Introduction to Seismic Interpretation* (2nd edn). Graham & Trotman Ltd, London.

Robinson, E. A. and Treitel, S. (1980). *Geophysical Signal Analysis*. Prentice Hall, Englewood Cliffs, New Jersey.

Sheriff, R. E. & Geldart, L. P. (1995). *Exploration Seismology* (2nd edn). Cambridge University Press, Cambridge, UK.

Stolt, R. H. and Benson, A. K. (1986). *Seismic Migration: Theory and Practice*. Geophysical Press.

Stone, D. J. (1994). *Designing Seismic Surveys in Two and Three Dimensions*. Society of Exploration Geophysicists, Geophysical Reference Series No. 5.

Tatham, R. H. & McCormack, M. D. (1991). *Multicomponent Seismology in Petroleum Exploration*. Society of Exploration Geophysicists, Investigations in Geophysics Series, Vol. 6.

Walden, A. T. & White, R. E. (1984). On errors of fit and accuracy in matching synthetic seismograms and seismic traces. *Geophysical Prospecting*, **32**, 871–91.

White, R. E. (1980). Partial coherence matching of synthetic seismograms with seismic traces. *Geophysical Prospecting*, **28**, 333–58.

Yilmaz, O. (1987). *Seismic Data Processing*. Society of Exploration Geophysicists, Investigations in Geophysics Series, Vol. 1.

3 Structural interpretation

This chapter is mainly about the most fundamental interpretation activity: making maps of horizons. Historically, it was the need for better maps of complex structural traps that was a key driver in the early adoption of 3-D seismic. Usually, however, it is not enough just to map the top of the reservoir. To understand how structures were formed and when, it is usually necessary to map a range of marker horizons above and below the target. Also, depth conversion will, in most cases, require the mapping of several horizons above the target level.

This chapter begins by considering how stratigraphic horizons encountered in wells can be tied to particular reflections on a seismic survey; this is an issue for all seismic interpretation, and is often easier on a 3-D survey because of the more nearly correct positioning of subsurface features. Having decided what to map, the interpreter is faced with the daunting problem of working with a huge number of seismic traces, all of which should ideally be taken into account. Happily, the power of computer workstations has increased faster than the quantity of traces waiting to be interpreted; the chapter continues with an explanation of how the tracking of reflecting horizons through a 3-D volume can be partly automated. This is not a matter only of mapping the horizon itself; in an area of even moderate structural complexity, it is the mapping of fault systems that will consume much of the interpretation effort. Semi-automated methods can help here too. For completeness, the chapter concludes with a brief discussion of how to convert a reflection–time horizon map into a depth map; the issues involved are the same as those with 2-D seismic, but the greater density of data in 3-D surveys may make the task easier in practice.

3.1 Well ties

One of the first steps in interpreting a seismic dataset is to establish the relationship between seismic reflections and stratigraphy. For structural mapping, it may be sufficient to establish approximate relationships (e.g. 'reflection X is near Base Cretaceous'), although for more detailed work on attributes, as described in chapter 5, it is usually necessary to be more precise and establish exactly how (for example) the top of a reservoir is expressed on the seismic section. Although some information can be obtained by relating reflectors to outcrop geology, by far the best source of stratigraphic information, wherever it is available, is well control. Often wells will have sonic (i.e. formation velocity) and formation density logs, at least over the intervals of commercial interest; from these it is possible to construct a *synthetic seismogram* showing the expected seismic response for comparison with the real seismic data. In addition, some wells will have *Vertical Seismic Profile (VSP)* data, obtained by shooting a surface seismic source into a downhole geophone, which has the potential to give a more precise tie between well and seismic data. In this section we shall discuss the use of both these types of data.

3.1.1   The synthetic seismogram

The basic idea is very simple. To a first approximation we can calculate the expected seismic response of the rock sequence encountered in the well by treating it as a one-dimensional problem. That is, we calculate the effect as though the interfaces in the subsurface are horizontal and the ray-paths are vertical, so that rays are normally incident on the interfaces. This is usually a reasonable first approximation, but means that we are ignoring the way that seismic response varies with angle of incidence, which will be discussed in chapter 5. In some cases, we may have to use the short-offset traces, rather than the full stack, for comparison with the calculated well response, to make sure that the approximation is valid. If we think of the subsurface as a number of layers, each with its own acoustic impedance A, then the reflection coefficient at the nth interface for P-waves at zero-offset is given by the formula

$$R_n = (A_{n+1} - A_n)/(A_{n+1} + A_n),$$

where A_n and A_{n+1} are the acoustic impedance above and below the interface; acoustic impedance is the product of density and seismic P-wave (sonic) velocity. The derivation of this formula can be found in Sheriff & Geldart (1995), for example; it is a simple particular case, valid for normal incidence, of the Zoeppritz relations, which describe how the reflection coefficient varies as a function of incidence angle.

Both density and sonic values are routinely logged as a function of depth in boreholes. Density is inferred from the intensity of back-scattered radiation from a downhole gamma-ray source; the amount of back-scatter is proportional to the electron density in the rock which is in turn proportional to the bulk density. Sonic velocity is determined from the travel-time of a pulse of high-frequency (e.g. 20 kHz) sound between a downhole source and downhole receivers; the sound travels as a refracted arrival in the borehole wall. Because of the methods employed, the values obtained for velocity and density are those in the formation close to the borehole wall, i.e. within a few tens of centimetres of the borehole. This may or may not be representative of the formation as a whole. It is possible that the well has drilled an anomalous feature, e.g. a calcareous concretion, at a particular level, which would give log readings quite different from those typical of the formation in general. Also, in permeable formations the zone adjacent to the borehole wall will be invaded by the drilling fluid, and this may alter the log response from the original formation values. In some cases, seismic velocity may vary significantly with frequency, a phenomenon usually called *dispersion*. In rocks of low porosity and permeability, the sonic log measured at 20 kHz may not be a reliable guide to velocities at seismic frequencies of 20 Hz.

Multiplying the sonic and density logs together will give us an acoustic impedance log. A typical display from a well synthetic software package is shown in fig. 3.1. On the left-hand side there is a scale marked in both depth and reflection (two-way) time (TWT); how we find the correspondence between depth and TWT is explained below. The values at the top of the scale show that a reflection time of zero corresponds to

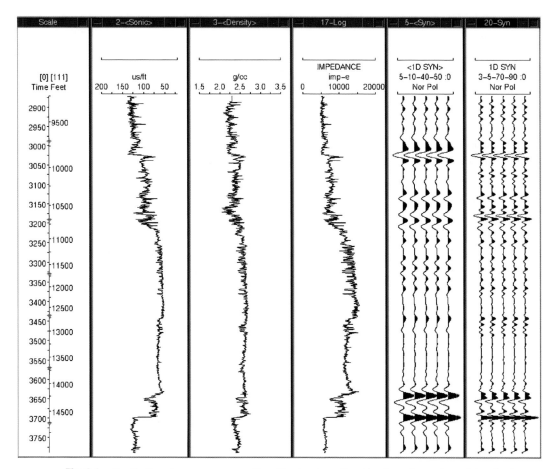

Fig. 3.1 Synthetic seismogram generation. Tracks show time/depth scale, sonic and density log, calculated impedance, and synthetic for two wavelets (5–10–40–50 Hz and 3–5–70–90 Hz, both zero-phase).

a depth of 111 ft; this is because in this offshore well, the datum for seismic times is sea-level but log depths are measured relative to a point on the drilling rig (the kelly bushing in this case) which is 111 ft above sea-level. There is scope for confusion here, particularly in the case of deviated wells where the depth scale may represent distance as measured along hole or alternatively may have been corrected to true vertical depth. Using erroneous datum values is a common cause of problems with well synthetics, and they need careful checking. The next three panels show the sonic and density logs, and the acoustic impedance log formed by multiplying them together. In routine logging practice, the sampling interval for sonic and density values will be half a foot in depth; thus the acoustic impedance log will typically show fine detail where thin interbeds of different lithologies are present. From this log it is possible to get an immediate impression of what causes the principal seismic reflections. They are often caused by sudden marked changes in impedance due to major changes in lithology, though they may also result from small low-amplitude impedance changes if they are cyclic at the right frequency to resonate with the seismic pulse (Anstey & O'Doherty, 2002).

The next step is to convert the acoustic impedance log, calculated from log data recorded as a function of depth, into a log as a function of (two-way) travel time. This is easy if we know the time–depth (T–Z) relation for the well, which can in principle be obtained by simply integrating the sonic log, though in practice two problems arise. One of them is that errors (for example, minor miscalibration of the sonic tool) tend to accumulate when the log is integrated over many thousands of feet. Another problem is that the sonic log is hardly ever run in the shallowest part of the hole. For these reasons, it is usual to calibrate the T–Z curve by means of some direct observations of travel-time from a surface source to a downhole geophone (*check shots*), e.g. at intervals of 500 ft along the entire borehole; the integrated sonic is then adjusted to match these control points. It is also possible to adjust the sonic log itself, and then to use this adjusted log to create the impedance values and the synthetic seismogram. This is often a bad idea; the piecewise adjustment of the sonic log tends to create a step change at each checkshot, and thus a spurious reflection. Obviously, it is possible to create a smooth adjustment to the sonic log that avoids this problem, but a simpler approach is to adjust only the T–Z curve and not the sonic log. A reflectivity curve is then calculated from the impedance using the formula given above. This reflectivity sequence is convolved with the wavelet thought to be present in the seismic data to generate the synthetic seismogram, the expected response of the logged interval, shown on the right-hand side of fig. 3.1.

There may be considerable uncertainty about the correct wavelet to use. The amplitude spectrum of the wavelet can be estimated from the seismic data, but in order to describe the wavelet completely the phase spectrum is also needed. This describes the relative shifts of the waveforms at each frequency (fig. 3.2). Two particular types of wavelet are often used: the minimum-phase and zero-phase wavelet. A minimum-phase wavelet is a causal wavelet, i.e. it has no amplitude before a definite start time.

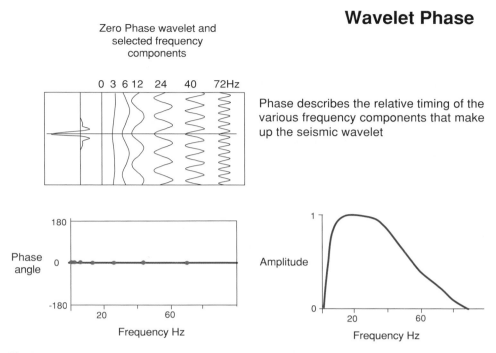

Wavelet Phase

Zero Phase wavelet and
selected frequency
components

Phase describes the relative timing of the
various frequency components that make
up the seismic wavelet

Fig. 3.2 The phase spectrum related to the timing relationship of the various frequency components in the wavelet, which in this example is zero-phase.

Of all the wavelets that have this property and have any particular amplitude spectrum, the minimum-phase wavelet is the one that has the most concentration of energy close to the start time. This does not necessarily mean that the leading loop is the largest; some minimum-phase wavelets have their greatest amplitude in the second loop. These wavelets are important because the actual source signature from explosives or air guns is close to minimum-phase. However, as noted in chapter 2, the wavelet is often converted to zero-phase during processing. This produces a wavelet like that shown in fig. 3.2, symmetrical about the zero-time and so with energy at negative times and not causal. This wavelet is preferred for interpretation because the strong central peak at time zero is easy to relate to the reflector concerned. In practice, the processor's attempt to convert the wavelet to zero-phase is rarely perfect, and mixed-phase wavelets are common.

The choice of wavelet can make a considerable difference to the appearance of synthetic traces (Neidell & Poggiagliolmi, 1977). There are several possible approaches. A simple method is to make synthetics with theoretical wavelets of different frequency content, both zero- and minimum-phase, and look for a good visual match to the real seismic. This may be adequate to identify the loops corresponding to key stratigraphic markers, so is a useful start to making structural maps. A totally different approach is to start from a measured source signature, and to calculate the way that it is modified by passing through the earth; however, the effect of the earth filter on wavelet phase may be hard to estimate. For detailed work involving measurement of loop amplitudes (see chapter 5), it may be best to estimate a wavelet from the data. This is easily done by using an algorithm that will calculate a wavelet that gives the best fit between the synthetic and the real data. The goodness of fit can be evaluated from the cross-correlation coefficient between the synthetic and the real seismic. However, the significance of high correlation depends on the length of the wavelet compared with the analysis window. If the wavelet is made long enough, then a perfect match can always be obtained, but such wavelets are often implausible (e.g. having high-amplitude oscillations) because they are trying to fit the noise in the real seismic as well as signal. A good match would be one with a high correlation over a long gate using a short wavelet. An empirical approach is to use a gate of 500 ms or more with a wavelet consisting of only 2–3 loops, but a more rigorous approach has been put forward by White (1980), who uses a statistical method to constrain wavelet length. With this type of approach, not only is the wavelet shape derived, but also the timing relative to time zero. This is important because some minimum-phase wavelets can look approximately symmetrical, and so roughly like a zero-phase wavelet, but the main loop is delayed from zero time. If we want to measure amplitudes on seismic data, it is important to measure the right loop, e.g. the one corresponding to the top of a reservoir.

In fig. 3.1, a zero-phase wavelet of frequency content 5–50 Hz has been used. The display repays close scrutiny. Firstly, it is a good idea to check the polarity of the display. As mentioned in chapter 1, there is often confusion about what the words 'normal' and 'reverse' mean as applied to the polarity of zero-phase seismic data. By looking at an isolated interface with a sharp impedance change, the polarity of the synthetic can be seen directly. Thus in fig. 3.1 there is a sharp increase in impedance at about 3020 ms which corresponds to a white trough (deflection to the left) in the synthetic. Next, it is easy to see where high-amplitude reflections are to be expected; these will be the easiest events to pick on the 3-D dataset to form a basis for structural mapping. A sharp impedance change gives the best response; the ramp-like impedance change at around 3200 ms causes only a moderate event, even though the total change in impedance is large. The fine structure within the impedance log is not represented at all in the synthetic. Detailed comparison for a particular target interval will show what hope there is of mapping, for example, the top and base of a reservoir from the seismic data. It is often useful to calculate synthetics for a range of high-frequency cutoffs, to see what bandwidth would be needed to reveal significant detail; if a moderate increase in resolution would improve the information significantly, it is worth considering additional seismic processing, inversion (chapter 6), or re-shooting the survey.

The comparison of the synthetic seismogram with traces extracted from the 3-D dataset around the well location is shown in fig. 3.3. In this case, there is a good visual match for the main events, but not for the weak events in the central part of the display. In such a case, it may be useful to calculate synthetic seismograms that include

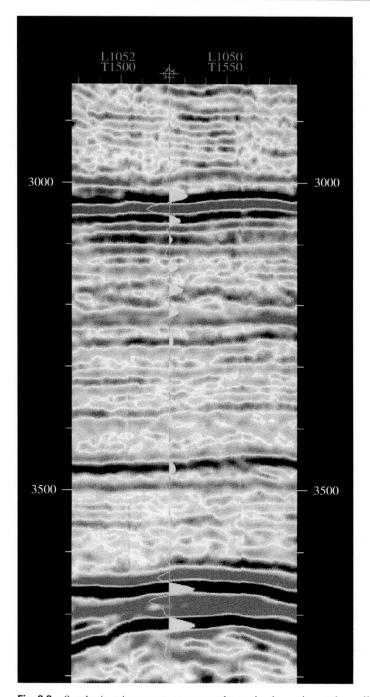

Fig. 3.3 Synthetic seismogram superposed on seismic section at the well location.

the effects of multiples. Ideally, seismic processing should have attenuated multiples to low levels relative to the primaries. If a better fit to the real seismic is observed when multiples are added into the well synthetic, then the interpreter needs to be on his guard; even if multiple energy is not obvious on a section view and does not cause problems for structural interpretation, it may still corrupt amplitude measurements on target reflectors. An advantage of having the dense data coverage of the 3-D survey is that it is possible to observe how correlation varies as a function of trace location. Ideally one would hope to see a bull's-eye pattern of high correlation centred about the well location, though the real world is not always so simple.

Sometimes it is easy to obtain a good match between well synthetic and seismic dataset; sometimes it is very difficult. There are many possible reasons for a mismatch.

Seismic survey problems	• incorrect zero-phasing (or other defective processing)
	• multiples
	• incorrect spatial location due to shot/receiver mispositioning or (more commonly) incorrect migration velocities
Synthetic seismogram defects	• defective logs
	• hydrocarbon effects
	• inadequate spatial sampling

In addition, there are the effects of amplitude variation with incidence angle; in principle this means that synthetic seismograms should be calculated for a range of angles and added together to simulate the stacked trace of the real data (see chapter 5).

It is important to understand the causes of an observed poor well to seismic tie. If the problem lies with the seismic dataset, similar problems will often be present in all wells within the survey area. Incorrect spatial location, if the result of mispositioning of sources and receivers, will probably mean that all ties will be improved by applying a constant lateral shift to the seismic trace locations. If the problem is incorrect migration of the surface seismic, then the tie points will be shifted updip or downdip by an amount related to the steepness of the local dip; thus flat-lying overburden might need little lateral shift, whereas deeper reflectors with dips of $30°$ might need lateral shifts of hundreds of metres.

Problems due to the synthetic seismogram will be different from one well to another. Various problems can arise with the wireline log data. They may need substantial editing to remove intervals of incorrect readings. Usually, the petrophysicist is interested mainly in obtaining good quality logs over the reservoir interval, which he or she will use to evaluate reservoir quality and hydrocarbon saturation; other intervals, especially shales, will not have received intense scrutiny for log quality at the time of acquisition. It is quite common to find data gaps and noisy intervals, owing for example to cycle skips on the sonic log where the automatic travel-time measurement system triggers on the wrong part of the signal pulse. Defective log intervals can be edited by removing noise spikes and by replacing erroneous data, either by plausible constant values or perhaps by values derived from another well. Over a particular interval, it may be possible to calculate values for one log from other logs, e.g. density from gamma-ray and sonic; the required relationships can be established from a nearby well where all the logs are of good quality. Whatever replacement method is used, care is needed to avoid introducing artificial sudden jumps in the sonic or density curves at top and bottom of the edited interval, as they would generate spurious reflections in the synthetic seismogram. Logs over hydrocarbon-bearing reservoirs should also be treated with great suspicion; if there is significant invasion of drilling fluid into the formation, either or both of the density and sonic logs may be recording values in a zone close to the borehole where the hydrocarbons have been partly swept away by the drilling fluid, whose properties are therefore not representative of the virgin formation. It is possible to estimate these effects using methods described in chapter 5, but the results are often unreliable because of uncertainty about the extent of the invasion and thus the magnitude of the effect on log response. Finally, as noted above, the logs sample the subsurface only within a few centimetres around the borehole, whereas surface seismic data respond to properties that are averaged laterally over at least several tens of metres. Thus, for example, a local calcareous concretion, which happened to be drilled through by a well, could show a marked effect on logs but have no seismic expression because of its limited lateral extent.

Even when the wireline log data are correct and representative of the formation, the approach described above may not result in a correct synthetic calculation. Implicitly, the method assumes that we can treat the propagation of the seismic wave through a 1-D earth model using ray theory. This is correct if the seismic wavelength is short compared with the layer thickness. If the wavelength is greater than about ten times the layer thickness (as will certainly be the case for surface seismic response modelled from closely sampled wireline data), then it is more appropriate to approximate the subsurface layering as an effective medium (Marion *et al.*, 1994). The effective medium velocity V_E is calculated as follows. Suppose we have a stack of thin layers in each of which there are log measurements of P velocity V_p, shear velocity V_s and density ρ. In each layer we determine the shear and bulk modulus (μ and K) from the equations

$$\mu = V_s^2 \rho$$

and

$$K = \rho \left(V_p^2 - \frac{4}{3} V_s^2 \right).$$

Over an interval (typically a quarter of the seismic wavelength), we then calculate the arithmetic density average and the harmonic average of μ and K. These average values are then used to calculate a mean value of V_p and V_s, using the same relations between the elastic moduli and velocities as before. This effective medium calculation is known as Backus averaging (Backus, 1962). The difference from the simple approach outlined previously depends on the velocity and density variation between the different layers. For shale–sand alternation, the difference is usually small; for shale–dolomite, it can be quite large, perhaps reaching as much as 20%.

It often happens that poor well ties cannot be definitely traced to a specific cause in the seismic or the log data. As pointed out by Ziolkowski *et al.* (1998), there is a fundamental problem: seismic traces are not in reality the simple convolution of a physically meaningful wavelet with the reflection coefficient series, as has been assumed above. In reality, the seismic data we collect are the response of a layered earth to a point source, including internal multiples and free-surface effects; conventional processing combines these records to produce stacked sections and attempts to remove multiples, diffractions, P–S conversions, and so on. It is not surprising if we then find that sometimes the wavelet that gives the best fit between the stacked seismic trace and the well synthetic varies from one well to another. Sometimes it may not be possible to have much confidence in the well tie, even after careful editing of the log data and consideration of all the other possible complications. At this point, it is useful to have a different line of evidence to help understand the cause of discrepancies between the well synthetic and the surface seismic, and this is where the VSP can help.

3.1.2   The VSP

Essentially, the technique is to record a surface source using downhole geophones; a general account has been given by, for example, Oristaglio (1985). The simplest geometry is the case of a vertical well with a seismic source at the wellhead (fig. 3.4(a), where the source is shown slightly separated from the wellhead for clarity). Recordings would be made of the source by a geophone at a series of downhole locations, e.g. at intervals of 50 ft over a vertical distance of 4000 ft to give records at 80 levels. The record for any particular level will contain both upgoing and downgoing waves; the former are reflections from horizons below the geophone, and the latter are the direct arrivals, as sketched in fig. 3.4(a). Both upgoing and downgoing arrivals will be contaminated by multiples due to reflections both above and below the geophone, but the upgoing waves that immediately follow the downgoing direct arrival have a useful property: they are free from multiple energy, because any multiple bounces in the ray-path would delay it and make it arrive significantly later than the direct ray. Also, the waveform of the direct arrival at a geophone tells us what the wavelet is at that particular depth. (The seismic wavelet changes slowly with depth owing to a progressive loss of the high frequencies.) This measured wavelet must also be the wavelet present in the reflections that immediately follow the direct arrival, because the travel path through the earth is nearly identical. Therefore, if we can separate the upgoing and downgoing arrivals, then the downgoing wavefield tells us the wavelet and allows us to calculate a filter operator to convert it to zero-phase. Applying the same operator to the upgoing

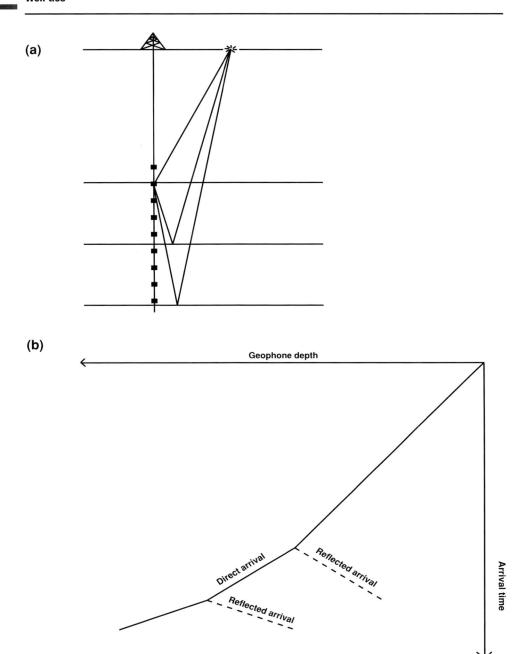

Fig. 3.4 (a) Schematic geometry of ray-paths from surface source to borehole geophones, for the simple case of a vertical well and small offset, shown exaggerated here for clarity; (b) schematic graph of arrival time against geophone depth.

wavefield produces zero-phase reflection data, with a more accurate control of the phase than can be achieved for surface seismic data.

Fortunately, it is quite straightforward to achieve the separation between up- and downwaves. As shown schematically in fig. 3.4(b), there is a difference in change of arrival time with depth between the up- and downgoing waves. When the geophone is located at the depth of a particular reflector, then the direct travel-time is the same as the reflected time to the event. As the geophone is moved up away from the reflector, the direct travel-time decreases and the reflection time increases. For the simple case of a vertical well with source at the wellhead, it is obvious that the decrease in travel-time for the direct arrival will be the same as the increase in travel-time for the reflection. In practice, it is fairly straightforward to measure the travel-times of the direct arrivals, though it is sometimes hard to identify the exact time at which the trace begins to deflect owing to the presence of noise. If traces are statically shifted by subtracting these times, then the direct arrivals will line up horizontally across a trace display; if the traces are shifted by adding the first arrival times (doubling the slope of the first arrival travel-time curve) then the upward travelling events will be horizontal (for horizontal reflectors, or nearly so if they are dipping). Filtering the first type of display to enhance laterally continuous events will result in an estimate of the downgoing wavefield, which can then be subtracted from the data to leave only the upgoing wavefield. Applying the second type of trace shift to these upwaves will then give us a display on which seismic reflectors are near-horizontal and can be enhanced by median filtering, which emphasises near-horizontal lineups in the dataset. A filter operator can also be applied to convert the wavelet (as measured in the downwaves) to zero-phase. It is then possible to form a *corridor stack* trace by stacking together the parts of the upwave dataset immediately following the direct arrival. This trace should then be zero-phase and free of multiples, and thus ideally suited for comparison with well synthetic and surface seismic. Since the VSP averages seismic response over a distance of a few tens of metres around the borehole, the problems of formation invasion and very small-scale lithological changes are not present. It is therefore usually helpful; the main problem is the rather low signal to noise ratio. Figure 3.5 shows an example display of deconvolved upwaves, after the traces have been shifted to make the reflected events line up horizontally. The corresponding enhanced (median-filtered) upwave display is shown in fig. 3.6. The median filter has brought out some consistent events that are scarcely visible in fig. 3.5; however, reliability of these events would need careful thought in a practical application.

Another advantage of the VSP is the ability to give good results in deviated wells, where synthetic seismograms are often unreliable, perhaps because anisotropy makes the sonic log readings (which measure velocity along the borehole) differ from the vertical seismic velocity in the formation; thus the calculated impedance contrasts are not those seen by a nearly vertically travelling ray. A useful VSP technique is the walk-above geometry, where the surface source is placed vertically above the geophone at a series of levels in the deviated hole. In this way, an image is produced of the zone

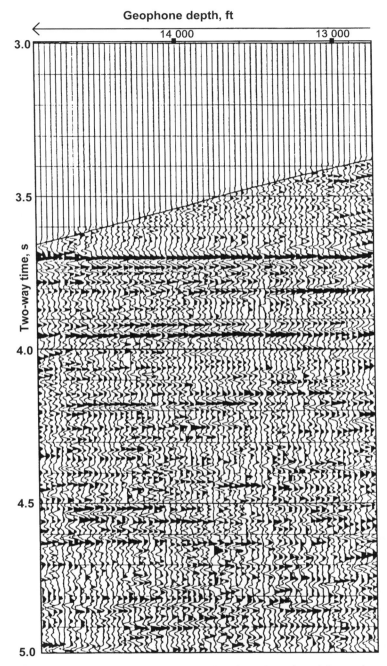

Fig. 3.5 Trace display for zero-offset VSP, after subtraction of downgoing waves, trace alignment of upgoing waves, and deconvolution.

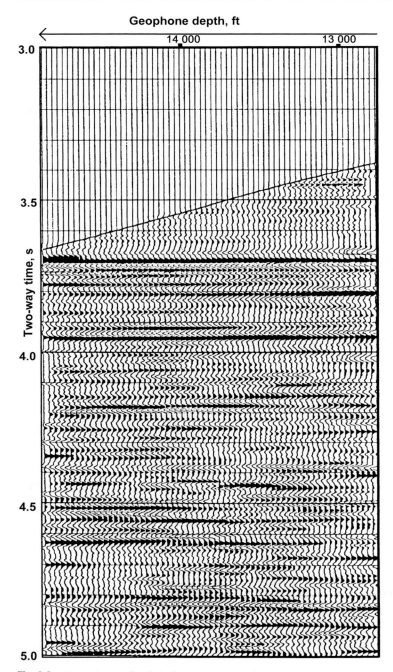

Fig. 3.6 Same data as fig. 3.5 after application of median filter to enhance horizontal alignments.

below the well bore, which can be compared directly with an arbitrary line selected from the surface seismic dataset. Note, however, that this technique does not work for a horizontal borehole; if there is no vertical separation between levels, it is not easy to separate the upgoing and downgoing wavefields, as there is no difference between them in the time–depth plot.

Although the VSP is often the best way to establish the tie between the surface seismic and the well information, it does have one disadvantage compared to the well synthetic. This is that it gives no real insight into how reflections are caused by changes in velocity and density values from one formation to another. This knowledge is vital, as we shall see in chapter 5, if we want to understand the likely causes of changes in seismic reflection amplitude or character from one part of the survey to another, which may allow us to predict lithology, reservoir quality or porefill.

3.2   Workstation interpretation

Having identified some horizons that are significant for understanding the geology and prospectivity of an area, the next task is to map them across the survey. Before carrying out any detailed work, it is useful to inspect the volume as a whole to get a general impression of structural and stratigraphic features of interest. This can be done by using the volume visualisation techniques discussed in chapter 7. Increasingly, interpretation is being carried out in this environment (Kidd, 1999). However, much detailed work is still carried out by picking the two-way reflection time to various horizons on some or all of the traces of the survey. These picked horizons are fundamental to the attribute measurement work discussed in chapter 5.

In the earliest days of 3-D survey, horizons were picked using methods carried over from interpretation of grids of 2-D lines. These were presented to the interpreter as a stack of paper prints; he would mark up the horizons of interest on a line through a well location and then follow them along the line to intersections with other lines, where the picks would be transferred to the crossing lines. By working round a loop of intersecting lines, it would be possible to get back to the starting point, where it could be checked that the interpretation was consistent around the loop. Interpretation would proceed by following the horizons round a series of such loops until they had been picked on all the lines (see McQuillin *et al.*, 1984, for a more detailed explanation with examples). The analogous method for 3-D data was to make paper displays of all the inlines and all the crosslines, with the idea of interpreting them in a similar way. However, it is easy to see that the effort needed for a full manual interpretation is prohibitive. Suppose we have a quite small rectangular survey consisting of 500 lines, each with 1000 traces. Then there would be 500 inline sections and 1000 crossline sections. The number of intersections to be checked would be 500 000. Assuming the data were easy to pick, it might be possible to verify the intersections at a rate of, say, 2000 per working day, so picking the entire survey would take a year of solid mechanical effort, with no time allowed for thinking about the meaning of the data. The practice therefore grew up of interpreting only a proportion of the data, say every 10th inline and crossline. Often this was sufficient for structural mapping; the benefit of having data correctly positioned in space after 3-D migration was secured. However, the fine detail that was present in the closely spaced data was lost.

The use of workstations for 3-D interpretation was therefore welcomed by inter-preters. They offered several advantages:

(i) the ability to view sections through the data in any orientation,

(ii) automatic book-keeping of manually picked horizons: picks made on one line would automatically be transferred to other lines or to map views,

(iii) semi-automated horizon picking,

(iv) calculation of pick attributes that can be used to extract additional information,

(v) ability to see the data volume in 3-D, not just as sections.

To achieve all this requires the use of fairly powerful workstations, and Appendix 1 describes some of the hardware and data management requirements. Each of the topics on the above list will now be addressed in turn.

3.2.1   Display capabilities

The 3-D seismic traces can be thought of as a volume of seismic amplitude values. In the example discussed in the previous section, there would be 500 000 traces arranged on a rectangular grid in map view, 500 inlines by 1000 crosslines. The two-way time on the vertical axis might range from 0 to 4000 ms, sampled at 4 ms, giving us 1000 samples on each trace. As shown in fig. 3.7, it is possible to view a range of different slices from this 'cube' of data. There are the obvious inlines and crosslines, but also horizontal slices (time slices), and vertical sections at any orientation through the volume. These 'arbitrary lines' do not have to be straight; they might, for example, be constructed to join up a number of well locations.

There are two possible modes of presenting seismic sections on the screen: as wiggle traces or as 'variable intensity' displays. In either case, there are limitations imposed by the screen resolution. This might, for instance, be 1024 by 1024 pixels. (A *pixel* is the smallest independently controllable element of a screen display. Software can specify the brightness and colour of each pixel on the screen but cannot achieve any higher (x, y) resolution than the pixel.) To get reasonable dynamic range on a wiggle trace display, the trace would need to extend over, say, 10 columns of pixels. If the traces do not overlap, this would imply that only 100 or so traces could be displayed at any one time. Traces can be allowed to overlap in order to view more of them, but even so a wiggle trace display will be limited to only a few hundred traces. This is suitable for detailed work (e.g. well ties or study of lateral changes in loop character), but makes it difficult to obtain an overview of the data. It is therefore often better to work in

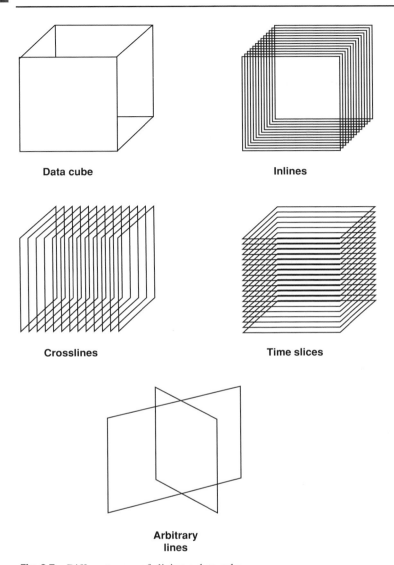

Data cube

Inlines

Crosslines

Time slices

Arbitrary
lines

Fig. 3.7   Different ways of slicing a data cube.

variable intensity mode. In this case, each trace is assigned one column of pixels, within which each pixel corresponds to a time sample; the pixels are colour-coded to show amplitude of the particular sample. Choice of colour-coding is under the interpreter's control, but popular choices are grey-scale (medium-grey for zero-amplitude, shading to black for large positive and white for large negative amplitudes) and red/blue or red/black dual polarity (white for zero-amplitude, shading to red for high negative amplitude and blue/black for high positive amplitude). It is useful to experiment with different colour bars; grey-scale often brings out subtle events (e.g. reflections oblique to the bedding, which may be noise or may carry genuine information about internal architecture from which depositional environments can be inferred) better than scales with violent colour contrasts, which tend to emphasise the continuity of events of a particular amplitude. Display scales are also under user control; it is possible to display e.g. only every other trace, or alternatively to interpolate extra traces so that real traces are separated by several interpolated ones, and similarly in the vertical axis. The interpreter needs to know the vertical exaggeration implied by his choice of scale factors; commonly, he wishes to see as much as possible of the lateral development of a geological feature at a vertical resolution adequate to pick a reflector accurately, and the result is often a display with a vertical exaggeration factor of e.g. 2–4. There is no harm in this so long as the interpreter is aware of it, but care is needed in picking faults on sections with high vertical exaggeration, where ramps can easily appear to be discontinuities.

Figures 3.8 and 3.9 show the same seismic section in variable intensity and wiggle trace forms respectively, and illustrate some of these points. The principal reflectors are easily interpretable on either display. The discontinuous events at the top of the section would most easily be picked on fig. 3.8; the greater trace spacing of the wiggle display comes close to aliasing these small reflector segments.

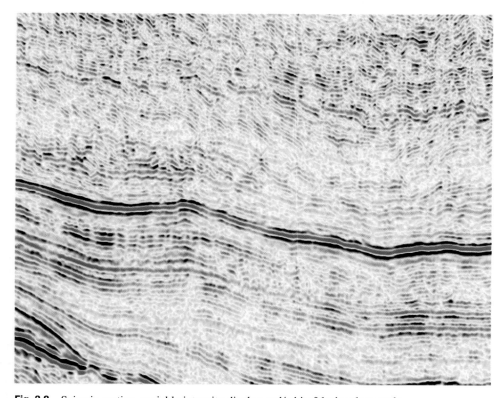

Fig. 3.8   Seismic section, variable intensity display, red/white/black colour scale.

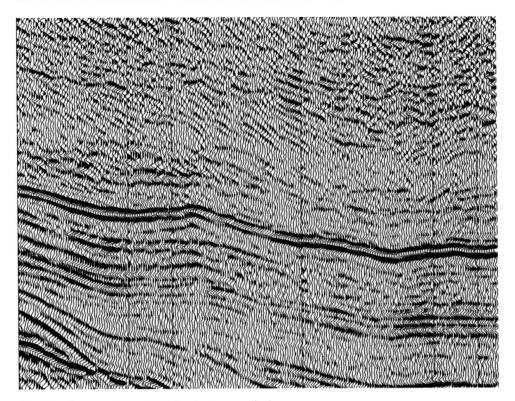

Fig. 3.9 Same section as fig. 3.8, wiggle trace display.

Another issue that the interpreter needs to be aware of is the effect of the limited dynamic range of the display and perhaps of the underlying data. The number of possible different amplitude levels that can have different colours assigned to them in a screen display is typically only 256 (8-bit resolution). This gives an adequate visual impression, but is much less than the dynamic range of the original seismic data. If the original data had a range of amplitude values from, say, $-13\,000$ to $+13\,000$ then the simplest way to convert it to the range from -127 to $+128$ required for the display would be to divide the seismic amplitudes by 100 and then round them to the nearest integer. However, the strongest amplitudes might be in a part of the dataset quite different from the one that we want to study; maybe we are interested in looking at subtle changes in reflection strength of a weak reflector, whose amplitude is only, say, $+1000$ in the original data. Scaling would reduce this to a value of around 10, and the effect of rounding to the nearest integer becomes serious: changes in reflection strength would be quantised in 10% jumps. The solution to this problem is to adjust the scaling of the display; the available display amplitude range of -127 to $+128$ is used for seismic amplitudes from $-C$ to $+C$, where C is a *clip level* set by the user. Any seismic amplitudes outside this range are simply set to -127 or $+128$ depending on the sign. By choosing the clip level correctly, it is possible to make best use of the screen display to show amplitude variation in the events that the interpreter is working with. Sometimes, however, instead of clipping being left to the interpreter's choice on final display, the decision is made for him or her during data loading from SEGY traces; the data are already clipped and scaled to the 8-bit range at the stage where they are made into a workstation seismic volume. This has advantages in reducing the volume of seismic data on the workstation disk and in decreasing the time taken to retrieve data from disk for display, because amplitudes are stored as 8-bit numbers rather than 16- or 32-bit. It will, however, cause problems if the interpreter wants to work with amplitudes of very strong events which have been clipped during data loading. It is therefore a good idea to establish the extent of clipping in the workstation dataset at an early stage; this is easily done by creating a display in which the extreme ends of the colour bar are set to a contrasting colour. An example is shown in fig. 3.10, where yellow has been substituted for red and black at the two ends of the colour bar. If this were a display of the data as loaded, without any additional clipping at the display stage, there would clearly be a problem in making amplitude measurements on the strong event in the centre of the picture. However, the data would still be quite satisfactory for general structural mapping.

An example of a time slice is shown in fig. 3.11. Time slices are useful for giving an instant map view, which may for instance make it simple to see the fault pattern. However, they are usually more difficult to understand than vertical sections, partly

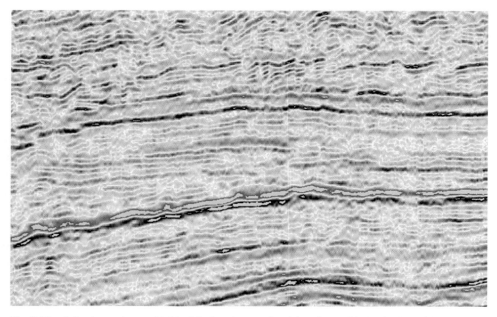

Fig. 3.10 Seismic section, red/white/black colour scale with yellow added at both positive and negative extreme ends of scale; yellow events have clipped amplitudes.

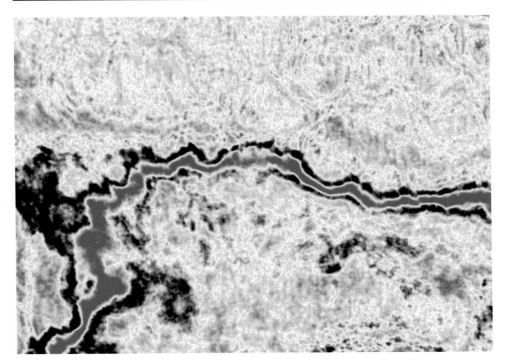

Fig. 3.11   Time slice display.

because reflectors dip through them more or less gradually and therefore do not have the distinctive character seen in vertical sections, and partly because there is no direct information on which way the reflectors are dipping. Close comparison with vertical sections is needed to identify the reflectors, and most workstation software has the ability to make composite displays of a time slice and an intersecting vertical section in order to achieve this easily. When looking for subtle stratigraphic features (e.g. channel systems) it is often helpful to make 'horizon slices' by slicing through the volume parallel to a picked horizon (which might for instance be the nearest easily picked loop above the level of interest). In these slices, the effect of structural dip is more or less removed, so a cleaner picture of the stratigraphic features is obtained.

3.2.2   Manual horizon picking

The process that is used for interpreting 2-D seismic using paper copies of sections is as follows. On the paper copies, horizons of interest are picked using pencils, generally with different colours for different horizons. Picks are transferred from line to line at intersections by folding one section and overlaying it on the other. Faults are marked up in the same way. When all the data have been picked, the TWT values to the horizons are read off the sections at a series of locations (e.g. every 10th CDP) and posted on a map in the correct location; fault intersections are marked in a similar way. The fault intersections are joined up between lines to establish the fault pattern, and horizon contours are constructed from the posted values.

All of these procedures have equivalents in 3-D workstation interpretation. Horizon picks are marked by digitising with a pointing device (usually a mouse) on a screen display of a section. This can be done using displays in any orientation, as explained in the previous paragraph. Once a horizon pick has been made on any particular trace of the 3-D data volume, it is available for display on any other section that includes that trace. For example, it is often best to start by picking along a series of composite lines that link the available wells together. It may then be best to interpret a few key dip lines across the structure; when these lines are displayed, the picks already made on the well traverses can be displayed automatically, to ensure consistent picks. A coarse grid of dip and strike lines might then be interpreted, which can later be infilled as much as required to define the features of interest. At each stage, the picks already made on intersecting lines can be displayed on the current section. Just as with paper data, this is a powerful check on consistency, and it is quite usual for picks to be deleted and reworked as the interpretation proceeds. (To make selective deletion possible, it is important to retain information on exactly what co-ordinates were used to construct particular composite sections through the data, if you are using anything more complicated than simple inlines and crosslines; all software allows you to store this information, but to do so is not always the default.)

Fault planes and their intersections with horizons are digitised from the screen display in a similar way. It is much easier to work with faults on lines crossing them approximately at right angles than on lines crossing them obliquely, where the fault plane crosses the bedding at a shallow angle. This is of course well known to the interpreter of 2-D data, where a line that crosses a significant fault obliquely will have a smeared image of the subsurface with substantial amounts of reflection energy coming from features out of the plane of section. In the case of 3-D data, the reflected energy has been repositioned so that the vertical section does not contain out-of-plane reflections, if the migration has been carried out correctly. Even so, it is difficult to recognise fault planes that do not make a high angle with the bedding, when projected on the line of section. This is because faults are almost invariably recognised from reflector terminations, as reflections from the fault plane itself are rare; the lineup of terminations is much easier to see on a dip section than a strike section (figs. 3.12 and 3.13). On the other hand, lines parallel to a fault may be very useful to investigate how one fault intersects with another, which may be crucial to the integrity of a fault-bounded structural closure.

While all this picking is going on, the software can continuously update a map display showing the horizon pick, by colouring in the traces on a basemap according to the TWT to the reflector. Fault intersections can be marked by special symbols on this map display. This makes it easy for the interpreter to keep track of what lines have been interpreted and of the emerging structural map. Usually interpretation workstations

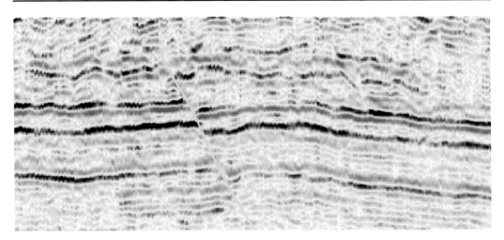

Fig. 3.12   Dip section through fault.

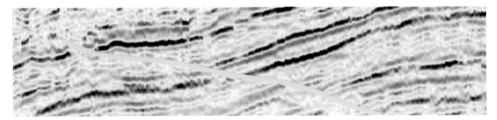

Fig. 3.13   Strike section through same fault as fig. 3.12.

have two display screens, so that one can be used for seismic sections and the other for maps.

One of the great benefits of 3-D seismic is that it is easier to establish fault patterns than with a grid of 2-D seismic. Semi-automated techniques will be discussed in section 3.2.4, but even with manual picks the density of the trace data avoids aliasing problems of the type shown in fig. 3.14. This shows a number of fault intersections on some parallel lines. If the faults all have similar general magnitude and appearance on the sections, it will not be obvious whether to join up the intersections as shown by the solid lines or as shown by the dashed lines. If these were data from a 3-D survey, the ability to view the crosslines would make it immediately clear which pattern is correct, although the problem is also much less likely to arise with 3-D data in the first place because the picked inlines are spaced much more closely than for a 2-D survey. It is usually fairly easy to correlate the faults in the map view of a horizon picked on a 3-D survey, and to draw polygons for the horizon cutout caused by the fault if it is normal (fig. 3.15). Reverse faults give rise to multi-valued horizons; if necessary the up- and downthrown sides can be mapped as separate surfaces, though in practice it is often sufficient to treat the faults as being vertical.

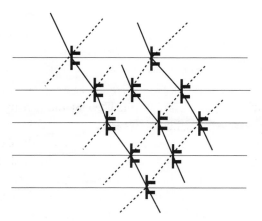

Fig. 3.14 Schematic map showing difficulties of establishing fault pattern from a grid of 2-D lines. Fault symbols show fault cuts visible on a series of east–west lines; they could be joined up by either the solid lines or the dashed lines (or some combination of them), leading to quite different fault maps.

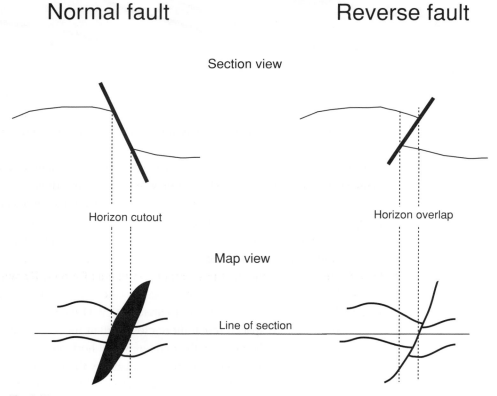

Fig. 3.15 Schematic section and map views of normal and reverse faults.

Once the fault polygons have been added to the map, the horizon surface can easily be gridded and contoured; if the horizon has been picked on a dense grid from the 3-D survey then a gridding step is not required, though it may still be useful as a way to apply smoothing and fill in small gaps in the interpretation. Unlike the case of a sparse 2-D dataset, where control of the gridding is a critical part of the interpretation, it is usually sufficient to use quite simple algorithms in view of the density of the picked horizon data. There are often problems, however, caused by minor inaccuracies in picking. Ideally a picked horizon should extend up to the picked fault plane and terminate exactly at the fault cut, on both the upthrown and downthrown sides, and the fault polygon on the map should match exactly the horizon cutout on the section. In practice, it would be extremely time-consuming to ensure total precision in picking horizons and faults on every line, and it is common to find some horizon values that belong to the upthrown side actually plotting on the downthrown side of the mapped polygon, and vice versa. This then causes anomalous grid and contour values near the fault, which is sometimes only aesthetically displeasing but which sometimes could lead to error in estimating the throw on a fault that might be critical to trap integrity. In such cases, the anomalous values have to be removed by detailed manual editing of the picks or of the grid values; alternatively, software is available to check the consistency of horizon and fault picks and to repair minor inconsistencies automatically.

3.2.3   Autotrackers

The snag with the procedure outlined in the previous section is that it is very time-consuming to interpret horizons manually across a large 3-D dataset, even with the aid of the automatic book-keeping provided by the interpretation software. A target horizon with complicated structuration may have to be tackled in this way, but in many cases the interpreter wants also to map simpler horizons; these may for instance be above the target zone and needed for depth conversion. In these cases, a semi-automated picking method is able to give good results in a fraction of the time needed for a full manual interpretation (Dorn, 1998).

The general idea is illustrated in fig. 3.16. Suppose we are trying to track a horizon that is a strong positive loop, and have identified it on one trace of a line, e.g. at a well. Then we might look for the pick on the next trace along the line as follows. Knowing that the horizon is at time T ms on the initial trace, we can look at a window from $T - \delta$ to $T + \delta$ ms on the next trace, choosing δ so that the correct pick should be within this window; this depends on the local dip of the reflector, and a reasonable value can easily be found by inspection of a section display. We assume for the moment that no faults are present. It is then possible to get the software to make a pick automatically within the window; the simplest approach would be to take the time of the largest positive amplitude. The process can then be extended to a third trace, using a window centred on the pick found on the second trace. In this way, the horizon could be picked along an

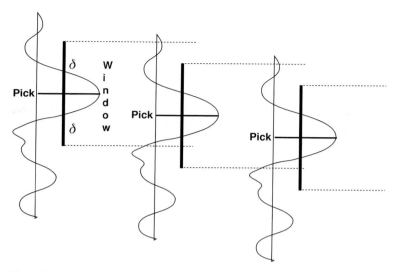

Fig. 3.16   Principle of a simple horizon autotracker.

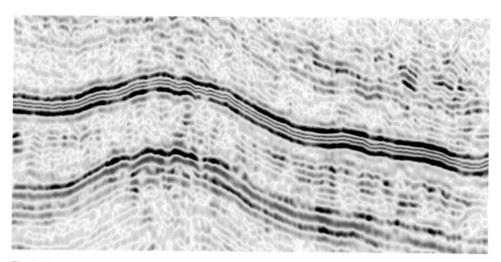

Fig. 3.17   Autotrack (in yellow) of a horizon with high amplitude and consistent character along the line.

entire line. Some protection against the effects of noise can be added by requiring that the amplitude of the maximum should be similar to that measured on the previous trace; if the amplitude change from one trace to the next exceeds a user-specified tolerance, the tracker simply stops and makes no further picks along the line. This type of tracking will work well only on a high-amplitude event, as shown in fig. 3.17. On a low-amplitude event, as shown in fig. 3.18, the results are far less satisfactory. A possible solution is to use a different type of tracking, which instead of just finding the maximum amplitude in

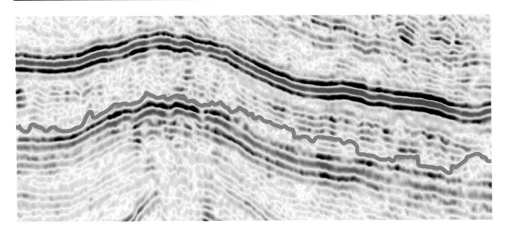

Fig. 3.18   Autotrack (in blue) of a low-amplitude horizon with no consistent character.

the window, tries to find a match for the trace shape by cross-correlating traces; this is computationally more demanding but will give better results on low-amplitude events.

The basic idea can obviously be extended to pick a negative maximum or a zero-crossing, or to take account of a pre-defined horizon dip when moving the picking window from one trace to the next. A more interesting extension is to apply the method across the entire 3-D volume. As soon as we attempt to do this, the algorithm needs to have some way of checking consistency of picks on a particular trace arrived at by different routes. This is analogous to the tying of loops in manual interpretation; if we start at a particular trace and pick a horizon until we reach another trace on a different inline and crossline, we want to get the same answer independent of the path through the data. Such a *volume autotracker* needs to check continuously for consistency as it picks away from an initial manually interpreted point, called a *seed point*. Differences in the way that the path through the dataset is controlled mean that different autotrackers will give different results on the same dataset in areas where the pick is difficult. In practice it is not possible to interpret a horizon across an entire 3-D dataset starting from an initial pick on just one trace. It is more realistic to use the autotracker as a sophisticated interpolation device, starting from a manual interpretation of a coarse grid of lines. Faults can be a particular problem. The autotracker may be able to find its own way correctly across minor faults, but there are often changes in reflection character across major faults, perhaps because of genuine changes in the geology. In such a case, it may be best to divide the area into panels between the major faults and autotrack each panel separately using its own seed picks. In any case, it is essential to review the results produced by the autotracker and to be prepared for several iterations of the process. The first pass will probably pick the simple areas correctly, but some areas will have to be erased and given an increased density of seed data until a satisfactory pick is achieved. Some software packages allow this erasure–reseeding–retracking process to take place under close user control, so that difficult areas can be picked with continuous control of the quality of the result.

The quality of an autotracked pick may be improved by pre-conditioning the seismic data using image processing techniques. One approach, structure-oriented filtering, has been introduced by Hoecker & Fehmers (2002). The idea is to stabilise reflections in the presence of noise, without smoothing over faults. The process consists of three steps: analysis of the raw data to determine the local orientation of the reflectors, edge detection to find reflection terminations, and smoothing of the data along the local orientation without filtering across the edges detected in the previous step. As well as removing noise, it is also possible to use such a filter to remove genuine but small-scale features of the data, such as very small faults or small-scale stratigraphic features. This opens up the possibility of an iterative approach to automated interpretation. In the first pass, all the fine detail is removed, permitting a rapid autotracking of the main horizons, and perhaps automatic fault tracking. This first result can then be fine-tuned by repeating the process on a dataset with less aggressive smoothing, stabilising the autotracking by using the result of the first pass as a seed grid. The process can be repeated through several cycles of iteration, until either the data have been interpreted to the required level of detail or the limit set by the noise in the dataset has been reached.

3.2.4   Attributes

A major advantage of workstation interpretation is that measurements of the seismic loop being picked are simple to calculate and store. The most obvious is loop amplitude, but loop width, average amplitude in a window below or above the horizon, and many others are commonly available. The ability to see these measurements in map view from densely sampled data is a key step in getting information from the seismic data about porefill (presence and type of hydrocarbons) and reservoir quality (porosity, net/gross, etc.). The way in which this can be done is the topic of chapter 5. Amplitude maps can also be the key to recognising stratigraphic features, e.g. channel systems. For accurate work, it may be important to know how the software calculates loop amplitude. Some early autotrackers would simply use the largest seismic amplitude seen at any of the (usually 4 ms) samples within the loop; since there is unlikely to be a sample exactly at the loop maximum, amplitudes were systematically underestimated. Modern autotrackers fit a curve to the amplitudes at the samples within the loop in order to estimate the true maximum value.

A different type of attribute is particularly important to structural mapping. It is possible to analyse both the picked horizons and the seismic trace data themselves to look for lateral discontinuities; we shall consider in this section those that are related to recognising faults, but they can also be used to aid geological interpretation in general.

The simplest approach involves calculation for a picked horizon of the local dip value and its azimuth (Dalley *et al.*, 1989). Figure 3.19(a) is a sketch map of a faulted

(a)

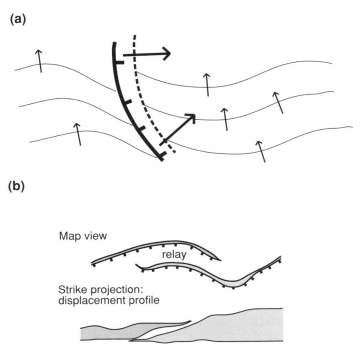

(b)

Fig. 3.19   (a) Schematic map to illustrate dip–azimuth calculation; (b) map view of relay ramp, after Davies *et al.* (1997).

horizon, showing contours and dip arrows. The unfaulted part of the horizon dips gently in a direction just west of north. Along the fault plane, provided the horizon is picked there, it will exhibit strong dips in an easterly or north-easterly direction. Most interpretation software provides the functionality to compute automatically the magnitude and direction (azimuth) of the local dip of a picked horizon, provided that it has been picked on every seismic trace, usually with an autotracker. Maps of these attributes will highlight the faults. This is a much easier way to arrive at a first-pass fault map than the time-consuming correlation of faults from one line to another in section view. Just as importantly, it will bring to the interpreter's attention faults of small apparent displacement that would probably not be recognised in section view but may be significant baffles to fluid flow in the reservoir. Note that not all faults will be visible on azimuth maps; if the fault happens to be parallel to the horizon contours locally, then there will be no change in the azimuth of the dip at the fault, though the fault will be obvious from its high dip values when a map of the magnitude of the dip is displayed. Dip maps are an effective way to study the individual fault segments comprising composite fault zones (Davies *et al.*, 1997). In extensional faulting, it is common for fault zones to consist of closely spaced overlapping fault segments; displacement is transferred across relay ramps between the terminations of the overlapping fault segments (fig. 3.19(b)). Sometimes these ramps are cut by faults that connect the terminations of the two main fault segments. Detailed mapping of the relay ramps may be important in understanding whether a fault will provide a seal to a prospective structure. Study of the way that displacement of a horizon varies along a fault is needed to assess sealing capacity (see chapter 4), and is also a useful check on the correctness of the interpretation; along a single fault, the displacement should increase smoothly from zero at the ends of the fault trace to a maximum in the centre.

It is also possible to calculate the local curvature of the picked horizon (Roberts, 2001). Faults appear as bipolar high-curvature anomalies, with high values of opposite sign produced where the fault plane intersects the horizon on its up- and downthrown sides. It may also be possible to relate curvature to fracture intensity, e.g. over salt swells. The main problem in using curvature (which is a problem for dip and azimuth calculation also) is to decide the length scale over which the attribute is calculated, distinguishing larger-scale structural features from small-scale features that might be sedimentological or might be noise (e.g. apparent reflector rugosity resulting from the autotracked pick wandering up and down within a broad loop of low signal to noise ratio).

Another approach is to make an illumination display of the picked horizon. The software calculates how the surface would look if seen from above when illuminated by a light source from a particular direction. Usually the source ('sun') direction is set to be near the horizontal ('low in the sky') so that subtle highs and lows in the surface are picked out by the contrast between the bright surfaces facing the sun and the dark shadows where surfaces face away from it. The effect is to emphasise those topographic lineations that in map view trend at right angles to the sun direction. Therefore, complete interpretation requires a number of illumination displays with the sun in different directions. Better still is to have interactive control of the sun position, with real-time updating of the screen display as the sun is moved around the map. In this way the user can choose sun positions to emphasise particular features of interest.

These methods depend on having an accurately picked horizon on a dense grid. If the horizon of interest is not easy to autotrack, the interpreter will have to do a good deal of editing and re-picking before he is able to use these tools. A different approach is to try to recognise faults as discontinuities in the seismic trace cube, without necessarily having any horizons picked at all. The basic idea is to calculate, over a limited time-gate, a measure of the similarity of a seismic trace to its neighbours (Bahorich & Farmer, 1995; an implementation is the subject of an Amoco patent). The calculated similarity value is posted in the seismic data cube at the centre of the trace window used to calculate it, and the process is repeated for every trace in the seismic volume and for every possible window start time; the result is therefore a complete data cube of similarity values. Faults are revealed as planes of low similarity; they are best seen in a horizontal section through the cube. The advantage over a simple time slice through the reflectivity data is that the faults will be visible whatever their orientation; on reflectivity slices, faults are often difficult to see where they run parallel to the strike of the bedding so that there

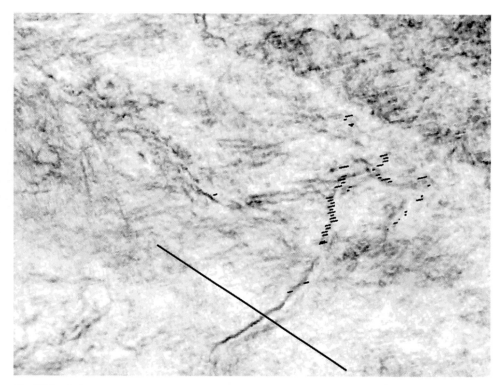

Fig. 3.20   Coherency map showing lineations due to faults; solid line is section of fig. 3.21.

are no obvious displacements of bedding lineaments. An example is shown in fig. 3.20, from the Lower Tertiary of the UK North Sea. A subdued colour-scale, such as the grey-scale used here, or a sepia scale, is often best for picking out subtle lineations. There are a number of lineations due to small faults. The bold black line marks the location of the seismic section shown in fig. 3.21. The small fault in the centre of the line is easily followed across the map view in fig. 3.20. Another example is shown in fig. 3.22, where a salt diapir pierces the horizon in the centre of the map and radial faults can be seen, especially in the south-eastern quadrant. Since the fault planes are surfaces of low coherence which are distinct from the higher coherence values within the 3-D volume generally, they can be visualised in 3-D from any perspective using the techniques discussed in chapter 7. They can also be autotracked using techniques similar to those used for horizons, though this has not yet become standard practice in the same way as horizon autotracking. The coherence cube methodology can also be used to reveal stratigraphic detail in the 3-D volume, such as channel/fan systems (see chapter 4).

Care is needed in interpreting all these attributes where there is significant coherent noise present in the seismic data. Interference of noise events (e.g. multiples) with real reflectors gives rise to discontinuities in reflectors that can be misinterpreted as faults

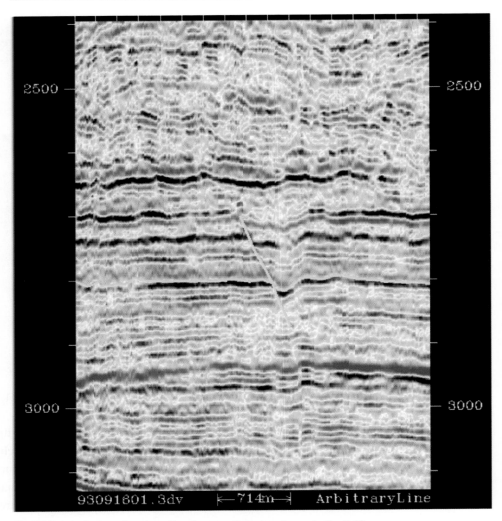

Fig. 3.21   Vertical section through coherency lineation shown in fig. 3.20.

(Hesthammer & Fossen, 1997). Careful study of seismic sections and comparison with well data are needed to avoid this pitfall.

3.2.5   Viewing data in 3-D

All the interpretation technology we have discussed so far works by presenting to the interpreter a 2-D section through the 3-D data cube. This may be a vertical section or a map view; as we have seen, the high density of trace data can be very helpful, particularly when making attribute maps. There is another possibility, however, which is to present the data to the interpreter as genuinely 3-D images. This is a rapidly

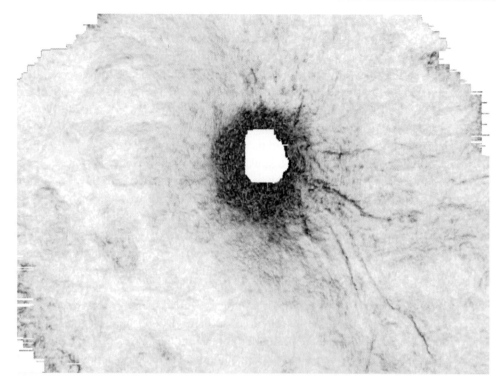

Fig. 3.22   Coherency map showing radial faults around a piercing salt diapir.

developing technique that is important enough to deserve a chapter of its own (see chapter 7).

3.3   Depth conversion

There is little that is specific to 3-D surveys so far as depth conversion is concerned; all the techniques in use are equally applicable to 2-D data. However, the density of data can make it much easier to incorporate seismic velocity information. For completeness, this section will cover briefly the general principles of vertical-stretch depth conversion, and the use of well velocity information and seismic velocity data; finally, we shall describe ways of dealing with complex overburden where lateral shifts are needed as well as vertical stretch.

3.3.1   Principles of vertical-stretch methods

If the seismic migration process has done a good job, then seismic energy is located at the right place in map view. If the migration was carried out using the wrong velocities, or using an algorithm that did not take proper account of overburden complexities, then there may be systematic lateral shifts of interpreted features (e.g. faults) from their true locations. These lateral shifts will be discussed in section 3.3.4; for the moment we assume that they are not a problem. Then all that is needed to convert the reflectors mapped in two-way time into depth is a knowledge of the seismic velocity in the subsurface. Sometimes, a quite detailed velocity model will already have been built by the seismic processors for migration purposes, but as we shall see these are not necessarily the best velocities to use for depth conversion.

To develop our ideas, it is useful to look at a real seismic section. A display of the entire section from surface to target level on a workstation is usually too poor in quality to use for detailed picking, owing to the limited vertical resolution, but is worth making to plan the strategy for depth conversion (fig. 3.23). In this example from the UK Central North Sea, the objective is at or just above the orange horizon, which is the top of the Ekofisk Formation. Two horizons have been picked in the overburden. They are levels at which there is substantial discontinuity in the curve of sonic velocity against depth at a nearby well (fig. 3.24). The yellow horizon is encountered at a level of about 1300 ms. Above this level, ignoring noise, velocity is nearly constant; at the horizon there is a slight decrease in velocity, and then velocity increases fairly steadily with depth to the top of the Sele Formation (green marker in fig. 3.23), at which point there is an increase in velocity and a more complicated velocity–depth trend which is only roughly approximated by a linear increase with depth. This suggests that a three-layer model would be suitable, with constant velocity in the top layer and velocity increasing with depth in the other two. A similar analysis needs to be carried out at an early stage of every interpretation, as it determines which overburden layers need to be picked to carry out the depth conversion. Sometimes, as here, only a few surfaces are needed to give a reasonable approximation. At other times quite a large number of surfaces may be needed, if there are large velocity jumps at a number of horizons; this might be the case if carbonates or evaporites are intercalated within a sand/shale sequence. To decide whether a given layer is worth including in the model, it is easy to calculate the error introduced at the well by treating it as part of an adjacent layer. To assess whether the error is important is harder, and depends on the detailed geometry of the structure being mapped; critical areas to look at will usually be the culmination and the possible spillpoint of any structural closure.

If the velocity within a layer is constant, it is obvious how to convert the two-way time thickness into a thickness in depth. If there is a velocity gradient with depth, we proceed as follows. Suppose we have a layer which extends from depth z_1 to depth z_2, at which the two-way times are t_1 and t_2 respectively, and that the velocity at any depth within the layer is given by

$$v = v_0 + kz.$$

In such a formula, v is often referred to as an *instantaneous velocity*; it describes the actual seismic velocity at a particular depth (or travel-time) and may be contrasted with

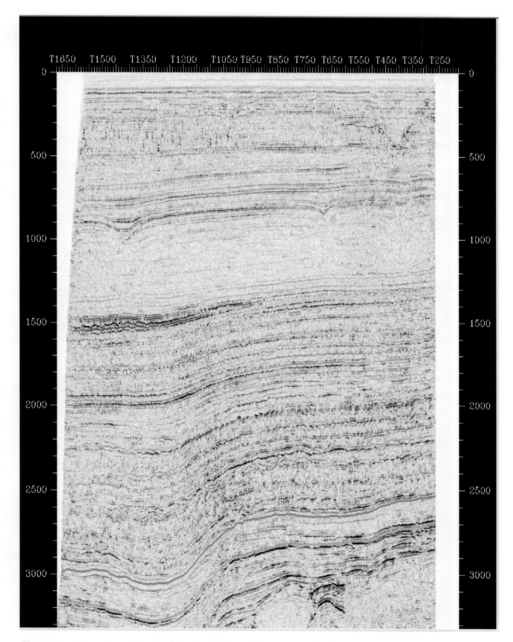

Fig. 3.23   Seismic section from surface to target (orange marker).

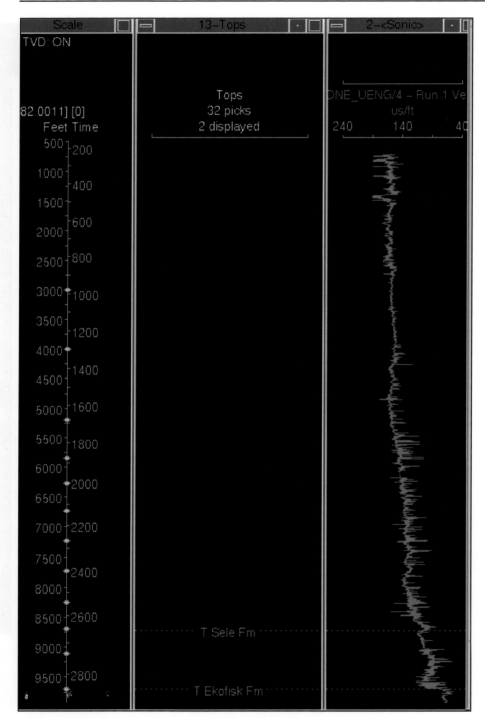

Fig. 3.24 Sonic log from near-surface to target (T[op] Ekofisk F[or]m[ation], orange marker in fig. 3.23).

the *average* velocity in the formation, which is simply the thickness in depth divided by the thickness in (one-way) time. For the seismic ray travelling vertically through the medium, the (two-way) travel-time t increases with depth z according to

$$\frac{\mathrm{d}z}{\mathrm{d}t} = \frac{(v_0 + kz)}{2}.$$

Therefore

$$\int_{t_1}^{t_2} \frac{\mathrm{d}t}{2} = \int_{z_1}^{z_2} \frac{\mathrm{d}z}{(v_0 + kz)}$$

so

$$\frac{k(t_2 - t_1)}{2} = \ln \frac{v_0 + kz_2}{v_0 + kz_1}$$

and thus

$$v_0 + kz_2 = (v_0 + kz_1) \cdot \mathrm{e}^{k(t_2 - t_1)/2}.$$

Therefore,

$$z_2 = \frac{(v_0 + kz_1) \cdot \mathrm{e}^{k(t_2 - t_1)/2} - v_0}{k}.$$

Whatever approach is adopted to splitting up the overburden into layers, a complication that often arises is that the depth map is distorted beneath zones where there is sharp lateral variation in the overburden, for instance under dipping fault planes. This should cause a kink in the time display of horizons below it, as shown in fig. 3.25. However, this is not observed in practice, except for pre-stack depth migrated data; if there is a significant velocity contrast between layer 1 and layer 2, there will be ray-bending at the fault plane, which will affect different offsets differently, leading to a smearing out of the expected kink in the stacked and migrated data. The quality of the reflected event will deteriorate under the fault plane, and the interpreter may have to pick through the noisy data by extrapolating the dip of the reflector as seen away from

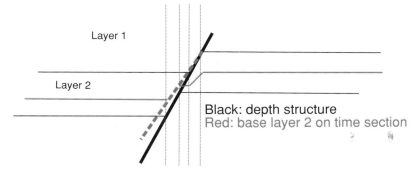

Fig. 3.25   Sketch of distortion of layer geometry on time section caused by a fault.

the fault. If the picked time horizon is smoothed out in this way, then vertical-stretch depth conversion will of course put a kink into the depth horizon under the fault plane. If this is a problem, then it may be an adequate approach to remove the distorted part of the horizon under the fault plane on the depth map, and replace it by extrapolating the horizon dip as seen outside the fault shadow. Smoothing or filtering the velocity field is another possible way to remove the distortion. If neither of these approaches is accurate enough (as might be the case for listric faults above a hydrocarbon accumulation or prospect, where the area affected by the fault shadow would be large), then pre-stack depth migration is needed.

A different approach to dealing with the effect of lateral variation in overburden velocity has been described by Armstrong *et al.* (2001). Look again at fig. 3.23; there are obvious channels in the near-surface, the largest of which is at the right-hand end of the section and extends down to a TWT of about 400 ms. There is some evidence on the section that the infill of this channel has a low seismic velocity; reflectors immediately underneath it are pushed down. Similarly, reflectors are pulled up by presumed high-velocity infill to the channel-like features visible at a TWT of 800–900 ms. Armstrong *et al.* proceed by measuring the push-down or pull-up on a reflector immediately below the anomalous feature, and use this information to simulate the effect on seismic lines acquired across it. As discussed in more detail in section 3.3.3, different source–receiver offsets are affected by the anomaly to differing extents, depending on the sum of the delays experienced at the two ends of the path. Simulation of CMP stacking of the modelled data then predicts the time distortion in the stacked data, whose effect can therefore be subtracted out of the horizon time map.

3.3.2   Use of well velocity information

Over the interval where sonic logs and checkshot data have been acquired, wells will have high-quality velocity information. If there is only one well, velocity values can be read from the log, or average velocities calculated for particular layers using the known two-way time (TWT) and depth at the top and bottom of the layer. As we saw above, usually there are small static shifts between well synthetic seismograms and real trace data; their effect can be removed from the depth map by using TWTs for the top and base of the layer taken from the picked seismic trace data rather than from the well sonic/checkshot information, when calculating average layer velocities. This is a reasonable approach if the static shifts do not vary much from one well to another; if they do, it would be better to identify and remove the cause of the shifts. If a $(v_0 + kz)$ model is being used, then values for v_0 and k can be found by fitting a line through the sonic log values plotted as a function of depth. If there are several wells, then the simplest possible approach is to average the values for each interval across all the wells. Another approach is to make maps of the velocities in each interval, or perhaps of v_0 and k values. Interpolation between the wells may be difficult, however. They may be few in number, and will usually have been drilled near the crests of anticlines; the interpreter, however, may need a reasonably accurate depth conversion of the intervening synclines, either to map the spill points of the anticlinal structures or to assess maturity of hydrocarbon source rocks. At first sight the $(v_0 + kz)$ model gives the required help, but this is not always the case. The simplest case is where the k factor reflects the effects of compaction, as unconsolidated sediments become more deeply buried over time. In such a case, the value found for v_0 may vary little from one well to another. This might be the case if all the rocks are currently at their greatest depth of burial. However, when rocks that once were deeply buried are later found near the surface after a period of uplift and erosion, the velocities usually remain close to what they were at the time of deepest burial. If there has been variation of the uplift from one well to another, then v_0 values will also vary. Simple interpolation of v_0 between the wells is valid only if uplift values can be similarly interpolated. Another possible complication is that the k factor may represent a change in velocity due to lithological effects, for example a consistent coarsening or fining upwards of a clastic sequence; k may then be quite similar from one well to another, but give no clue about the effect of depth on velocity. Rather than deriving k from the sonic log, it may therefore be preferable, where several wells are available, to determine a compaction trend by plotting the average velocity in each formation against midpoint depth. The gradient of this line (κ) is not the same thing as the k value for instantaneous velocity unless the interval is thin (time thickness much less than $1/k$). For thick intervals, with the notation of the previous section, we would find:

$$\frac{(z_2 - z_1)}{(t_2 - t_1)/2} = v_0 + \kappa \left(\frac{z_2 + z_1}{2} \right)$$

so that

$$z_2(1 - \kappa(t_2 - t_1)/4) = z_1 + (v_0 + \kappa z_1/2)(t_2 - t_1)/2$$

and

$$z_2 = \frac{4z_1 + (2v_0 + \kappa z_1)(t_2 - t_1)}{4 - \kappa(t_2 - t_1)}.$$

However the velocity maps are calculated, it is usually a requirement that the final depth map should match the formation tops in the wells. This will always be the case if the velocity derivation methodology honours the well data exactly, as can easily be done if maps are being made of the velocity in each layer. However, if some or all layers are depth-converted using constant parameter values (e.g. the average for the velocity in the layer, across all the wells), then there will be discrepancies between the depth map and the true well depths. If they are large, the method of depth conversion used needs to be revisited; if they are small, the usual practice is to grid up the mistie values and apply them as a correction across the whole map. The gridding algorithm needs to be chosen so that it will not produce unreasonable values outside the area of well control, as might be the case if the gradients between wells are extrapolated beyond them. It is not usually possible to understand exactly what controls these residual discrepancies, so the scope for intelligent contouring taking account of geological trends is limited; this is why the residuals should be quite small before this step is taken.

Quite often, well data give us information at so few points that it would be useful to bring some extra information into play to interpolate between them. To get velocity information across the whole of the seismic survey, it is natural to turn to the velocity fields derived during the processing of the seismic data.

3.3.3   Use of seismic velocities

During the course of seismic processing, a densely sampled velocity field is generated in order to stack and to migrate the data. It is often assumed that stacking velocities are root-mean-square (rms) average velocities from the surface down to the reflector concerned. It was shown by Taner & Koehler (1969) that for a reflector at the base of n uniform horizontal layers, the reflection time T_x corresponding to a source–receiver distance x is given by

$$T_x^2 = T_0^2 + \frac{x^2}{V_{\text{rms}}^2} + C_3 x^4 + C_4 x^6 + \cdots,$$

where the coefficients C are functions of the thicknesses and velocities of the n layers and V_{rms} is the rms velocity along the zero-offset trajectory defined by

$$V_{\text{rms}}^2 = \frac{\sum\limits_{k=1}^{n} v_k^2 t_k}{T_0}.$$

Stacking velocities V_{st} are usually calculated by methods that assume a hyperbolic time–distance relationship, i.e. that fit a relationship of the form

$$T_x^2 = T_0^2 + \frac{x^2}{V_{\text{st}}^2}$$

to the travel-time versus offset data. The stacking velocities are therefore only an approximation to the rms average velocity from the surface to the reflector concerned. However, there are other reasons why the velocities that give the strongest stack amplitudes and the most sharply focussed reflections are only loosely related to the actual seismic velocities in the real earth. Al-Chalabi (1994) has provided a useful summary of them. The most serious effects on stacking velocities are due to statics, structure and anisotropy.

Statics effects arise when the survey is shot over a near-surface velocity anomaly (Al-Chalabi, 1979). The geometry is shown in fig. 3.26 for the case of a model involving a step across which a near-surface delay is generated. When the CMP location is at A, only the outer traces of the CMP gather experience the delay; the best-fit hyperbola

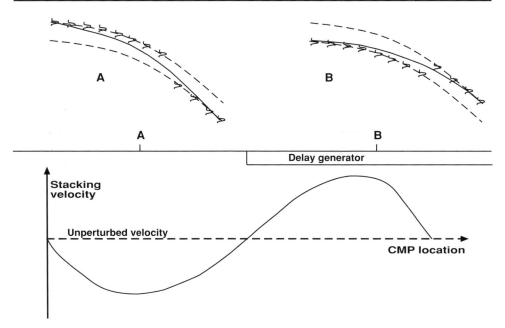

Fig. 3.26   Sketch of effect of near-surface static-type delays on stacking velocity (after Al-Chalabi, 1979).

across all the data will be steeper than the hyperbolae followed by the inner or outer trace arrivals separately, corresponding to a lower velocity. When the CMP location is at B, most of the rays see a double delay, once on the shot side and once on the receiver side; the outer traces experience only a single delay. The best-fit hyperbola across all the traces is then flatter (higher-velocity) than would be found for the unperturbed traces. As shown at the bottom of the figure, the result is an approximately antisymmetric variation of stacking velocity, with a wavelength equal to the spread length. The effect can be large, with oscillations up to 15% of the average velocity to the reflector. If the velocity anomaly is at depth, rather than at the surface, both the width of the stacking velocity response and its amplitude are reduced. In general, lateral variations in the stacking velocity field with a wavelength less than the spread length (maximum source to receiver distance) are not to be trusted; if a sufficiently large number of stacking velocity data are available, the spurious effects can be largely removed by smoothing the data.

The effect of structure arises in several ways. Dip can have a significant effect on velocity estimates. Often dip-independent velocities will be available if DMO has been applied during processing; if not, a correction can be made (Levin, 1971). However, to use the seismic velocities for depth conversion we usually want to calculate interval velocities in each individual layer. This is done by means of the Dix formula (Dix, 1955):

$$V_{\text{int}} = \left(\frac{(V_b^2 T_b - V_a^2 T_a)}{(T_b - T_a)} \right)^{1/2},$$

where V_{int} is the interval velocity in a layer with rms velocities V_a and V_b to its top and base, and T_a and T_b are the corresponding normal incidence times. The formula does not take account of ray-bending effects, and gives incorrect results for dipping interfaces (with dips larger than about 7° for cases modelled by Al-Chalabi). Reflector curvature also biases velocities, and is not easy to correct for.

Anisotropy arises as an issue because the velocities determined from seismic processing are, broadly speaking, horizontal velocities through the ground; for depth conversion, we of course need to have vertical velocities. Many rocks, however, exhibit anisotropy, with horizontal velocities larger than vertical ones. This may be intrinsic to the rock or an effect of small-scale interbedding of faster and slower lithologies. It is possible to measure anisotropy (e.g. from long-offset VSP data; Armstrong *et al.*, 1995), but often no direct measurements are available and it has to be inferred from the comparison of seismic and well data.

In general, migration velocities are closer to true velocities in the ground than are stacking velocities, because of the removal of structure effects and much of the statics effect; anisotropy remains a serious factor, however, and it is not possible to use a migration velocity directly for accurate conversion from time to depth. However, if we are looking for a way to interpolate velocities between wells, migration velocities can be useful. We can compare the actual well velocities in a particular formation with the migration velocities measured at the well locations, and so estimate a correction factor; if anisotropy does not vary much laterally within the formation, then it should be possible to use a single correction factor for it across the entire area.

All these comments apply as much to 2-D as to 3-D seismic data. The main benefit of 3-D data is that the velocity field will have been densely sampled in space. It is likely to be of better quality than velocities derived from isolated 2-D lines because of the opportunity to spot mistakes by plotting out sections through the velocity cube (e.g. horizontal slices), and because it is densely sampled it can easily be smoothed to remove the effects of statics. However, it is still only really suitable as a way of interpolating between wells.

3.3.4   Lateral shifts

Sometimes the accurate lateral positioning of events in the seismic dataset is very important. An example might be the case of planning a well to drill into a fault block on the upthrown side of a major fault. There may be a need to drill as close to the fault as possible, perhaps to ensure maximum drainage of a reservoir compartment, but it will be crucial to drill on the correct side of the fault and not accidentally on the downthrown side, which might be outside the hydrocarbon accumulation altogether.

Accurate lateral positioning depends mainly on the quality of the seismic migration process; for modern surveys, any uncertainty in the surface positions of shots and receivers is negligible by comparison. It is important to realise that migration may

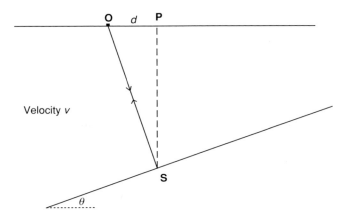

Fig. 3.27 Zero-offset reflection ray-path for a dipping layer.

shift reflectors by large distances laterally. Consider the very simple case shown in fig. 3.27, where a single dipping reflector is overlain by constant-velocity overburden. An identifiable point S on the reflector (perhaps a small fault) will be imaged on the stack section below the point O where the zero-offset ray intersects the surface; OS is perpendicular to the reflector. The migration process has to shift the image laterally to the true location below P, over a distance d. Then

$$d = \text{OS} \sin \theta$$

and if the two-way travel time for the zero-offset ray is t, then

$$d = \frac{vt}{2} \sin \theta.$$

Of course, we do not directly observe the true dip in depth, but rather the dip on the (unmigrated) time section, the rate of change of t with d, which is given by

$$\frac{2 \sin \theta}{v}.$$

Call this quantity q. Then

$$d = \frac{vt}{2} \cdot \frac{qv}{2} = \frac{v^2 t}{4} \cdot q.$$

The error δ in d due to an error δv in v is then given by

$$\delta = 2v\delta v \cdot \frac{qt}{4},$$

or

$$\frac{\delta}{d} = \frac{2\delta v}{v}.$$

For example, suppose the overburden velocity is 3000 m/s. For an event at 2 s two-way time with a dip of 15°, the lateral shift would be 3000sin 15°, or 776 m. A 2% error in estimating the velocity v would result in a 4% error in this shift, or about 31 m. This is not likely to cause problems. However, for a reflector dipping at 25° at a time of 4 s, the shift would be 2535 m and a 4% error in this would be 101 m, enough to cause serious concern when planning a well. In such a case, careful investigation of the migration velocity is needed to establish its likely accuracy.

Complications arise as soon as the overburden shows significant velocity variation, particularly if there are rapid velocity changes laterally. The effect of ray-bending in the overburden then has to be taken into account. Time migration algorithms assume hyperbolic moveout, and account for lateral velocity variation by varying the shape of the hyperbola with map location; this is satisfactory only if the velocity structure does not vary laterally across a CMP gather. The technically correct approach in the case of rapid lateral variation is pre-stack depth migration. This is, however, time-consuming and expensive, because of the effort needed to build a correct 3-D velocity model; if the model is incorrect, the migrated image may be worse than that from a simple time migration. Various methods have therefore been suggested to apply corrections for lateral shift to time-migrated data. One method is the use of image rays. The basic idea (Hubral, 1977) of the image ray is that it starts vertically downwards from a point at the surface, and propagates through the subsurface refracting at all velocity boundaries until the travel time is used up. The corrected horizons are positioned at the end point of the image rays. However, image rays will correct for ray-bending only in the case where the target horizon has zero time-dip; in other cases the lateral displacement derived by this method will be incorrect, because the overburden sampled by the image ray is different from that seen by the actual physical rays reflected from the dipping surface (Calvert, 2002).

References

Al-Chalabi, M. (1979). Velocity determination from seismic reflection data. In: *Developments in Geophysical Exploration methods – I* (ed. A. A. Fitch), pp. 1–68. Applied Science Publishers, Barking.

(1994). Seismic velocities – a critique. *First Break*, **12**, 589–96.

Anstey, N. A. & O'Doherty, R. F. (2002). Cycles, layers and reflections. *The Leading Edge*, **21**, 44–51.

Armstrong, P. N. Chmela, W. & Leaney, W. S. (1995). AVO calibration using borehole data. *First Break*, **13**, 319–28.

Armstrong, T., McAteer, J. & Connolly, P. (2001). Removal of overburden velocity anomaly effects for depth conversion. *Geophysical Prospecting*, **49**, 79–99.

Backus, G. E. (1962). Long-wave elastic anisotropy produced by horizontal layering. *J. Geophys. Res.*, **67**, 4427–40.

Bahorich, M. & Farmer, S. (1995). 3-D seismic discontinuity for faults and stratigraphic features: the coherence cube. *The Leading Edge*, **14**, 1053–8.

Calvert, R. (2002). Image rays and the old myth about correcting time migrated positioning. *First Break*, **20**, 715–16.

Dalley, R. M., Gevers, E. C. A., Stampfli, G. M., Davies, D. J., Gastaldi, C. N., Ruijtenberg, P. A. & Vermeer, G. J. O. (1989). Dip and azimuth displays for 3D seismic interpretation. *First Break*, **7**, 86–95.

Davies, R. K., Crawford, M., Dula, W. F., Cole, M. J. & Dorn, G. A. (1997). Outcrop interpretation of seismic-scale normal faults in southern Oregon: description of structural styles and evaluation of subsurface interpretation methods. *The Leading Edge*, **16**, 1135–43.

Dix, C. H. (1955). Seismic velocities from surface measurements. *Geophysics*, **20**, 68–86.

Dorn, G. A. (1998). Modern 3-D seismic interpretation. *The Leading Edge*, **17**, 1262–73.

Hesthammer, J. & Fossen, H. (1997). The influence of seismic noise in structural interpretation of seismic attribute maps. *First Break*, **15**, 209–19.

Hoecker, C. & Fehmers, G. (2002). Fast structural interpretation with structure-oriented filtering. *The Leading Edge*, **21**, 238–43.

Hubral, P. (1977). Time migration – some ray theoretical aspects. *Geophysical Prospecting*, **25**, 738–45.

Kidd, G. D. (1999). Fundamentals of 3-D seismic volume visualization. *The Leading Edge*, **18**, 702–12.

Levin, F. K. (1971). Apparent velocity from dipping interface reflections. *Geophysics*, **36**, 510–16.

Marion, D., Mukerji, T. & Mavko, G. (1994). Scale effects on velocity dispersion: from ray to effective medium theories in stratified media. *Geophysics*, **59**, 1613–19.

McQuillin, R., Bacon, M. & Barclay, W. (1984). *An Introduction to Seismic Interpretation*. Graham & Trotman, London.

Neidell, N. S. & Poggiagliolmi, E. (1977). Stratigraphic modelling and interpretation – geophysical principles and techniques. In *Seismic Stratigraphy – Applications to Hydrocarbon Exploration* (ed. C. E. Payton), AAPG memoir **26**, pp. 389–416.

Oristaglio, M. L. (1985). A guide to current uses of vertical seismic profiles. *Geophysics*, **50**, 2473–9.

Roberts, A. (2001). Curvature attributes and their application to 3D interpreted horizons. *First Break*, **19**, 85–100.

Sheriff, R. E. & Geldart, L. P. (1995). *Exploration Seismology*. Cambridge University Press, Cambridge, UK.

Taner, M. H. & Koehler, F. (1969). Velocity spectra – digital computer derivation and applications of velocity functions. *Geophysics*, **34**, 859–81.

White, R. E. (1980). Partial coherence matching of synthetic seismograms with seismic traces. *Geophysical Prospecting*, **28**, 333–58.

Ziolkowski, A., Underhill, J. R. & Johnston, R. G. K. (1998). Wavelets, well ties, and the search for subtle stratigraphic traps. *Geophysics*, **63**, 297–313.

4 Geological interpretation

All seismic interpretation is of course directed toward geological understanding of the subsurface. In the previous chapter, the objective of the interpreter was to make maps of surfaces, mainly in order to delineate traps by mapping the top of a reservoir. However, how does he or she recognise where the reservoirs are likely to be in an undrilled area? What reflectors are most likely to be the top of a reservoir body? If there are some well data available, perhaps reservoirs have already been encountered, but what is their lateral extent likely to be? What lateral changes in reservoir quality are likely, and how should they be related to changes in seismic appearance? These questions are of course just as relevant for 2-D seismic as for 3-D, but the dense data provided by 3-D seismic offers more scope for defining the external geometry and internal architecture of reservoir bodies. The detailed map view derived from 3-D seismic is often more instructive than an individual section can be.

Before embarking on a more detailed discussion, it is important to understand the limitations on achievable seismic resolution; this is discussed in section 4.1. The principles of seismic stratigraphy are briefly explained in section 4.2, including the recognition of seismic facies. Some tools to allow the interpreter to look for the expression of different sedimentary facies are described in section 4.3, and some examples of the results presented in section 4.4.

The structural geologist also has of course an input to make to 3-D interpretation. The need for validation of fault patterns is less than in the case of 2-D surveys, where aliasing of fault patterns can be a major issue. However, understanding of fault systems may be critical to understanding whether faults will form effective lateral seals. These topics are discussed in section 4.5.

4.1 Seismic resolution

Both vertical and horizontal resolution of seismic data are limited, and this imposes limits on what geologically significant features can actually be recognised on seismic data. Vertical resolution is determined by the seismic source signal and the way it is filtered by the earth. For example, the signature of a typical marine air-gun array has

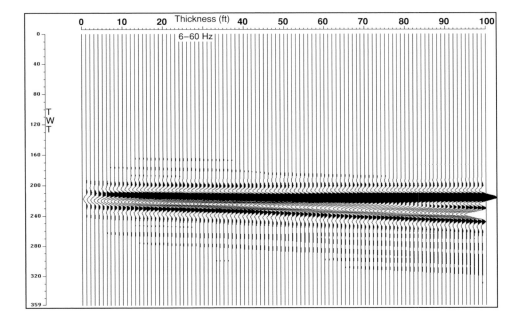

(a)

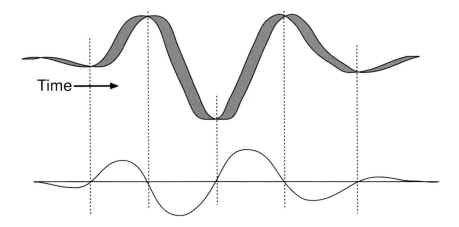

(b)

Fig. 4.1 (a) Wedge model for bandwidth 6–60 Hz; (b) model for calculating thin bed response from the difference of identical wavelets slightly displaced in time, redrawn after Widess (1973) with permission of the SEG.

frequencies in the range 8–150 Hz; the upper frequency limit will be reduced as the seismic signal propagates through the earth, perhaps to about 50 Hz at a TWT of 2 s. The effect of finite bandwidth can be studied using a simple model (fig. 4.1(a)). This shows the zero-offset response to a wedge of material increasing in thickness from zero to 100 ft in 1 ft increments. The reflection coefficient at the top of the wedge is −0.15 and that at the base is +0.185. The material of the wedge is therefore significantly softer than the material above or below it, and the material above it is slightly softer than that below it. These values are based on those for a wedge of porous gas-filled sand encased in Tertiary shales in the UK Central North Sea. Figure 4.1(a) shows the calculated seismic response for a zero-phase wavelet of bandwidth 6–60 Hz. The polarity of the display is that a black peak marks a transition downwards to an acoustically softer material. Where the sand is absent, at the left-hand end, there is a weak white trough due to the impedance difference between the shales above and below the sand level. At the right-hand end, the top of the sand is marked by a strong black loop and the base by a strong white loop. There are small-amplitude wiggles between, above and below these reflectors, caused by minor oscillations in the wavelet, but it would clearly be possible to pick the strong loops at top and base sand accurately, and measure the TWT interval between them to determine sand thickness. As the sand becomes thinner, however, the separation between the top and base loops reaches a nearly constant value at a thickness of about 40 ft. The point at which this happens is often called the *tuning thickness*. After this, the separation remains nearly constant, and further decrease in sand thickness causes the amplitude to decrease. This is the result of *interference* between the reflections at the top and base of the sand; the reflections from the top and base overlap and, being of opposite polarity, partly cancel one another. Below 40 ft thickness, the top and base sand are not visible as separate events. It is very important to take this into account when estimating reservoir volumes in thin sands; using the isopach between top and base seismic reflectors will grossly overestimate the volume.

A method to calculate thicknesses for thin sands, below the tuning thickness, was discussed by Widess (1973), using a simple model where the reflection coefficients are the same at the top and base of the bed. As shown in fig. 4.1(b), the resulting signal is the sum of the reflections from the top and base of the bed, which are of course of opposite polarity; it is therefore the difference between two identical wavelets slightly displaced in time. When the bed is very thin, the character of the reflection is that of the time derivative of the incident wavelet. Widess showed that the character of the composite reflection is unchanging for beds whose thickness is less than about $\lambda/8$, where λ is the wavelength in the bed material corresponding to the predominant period of the wavelet. For beds thinner than this, reflection amplitude is given by $4\pi Ab/\lambda$, where A is the amplitude that would be obtained from the top of a very thick bed (i.e. with no interference effect), and b is the thickness of the bed. Thus the amplitude is proportional to bed thickness for these thin beds, and this can be used to predict bed thickness from seismic amplitude if the data are calibrated (e.g. to a well) and if we can assume that all lateral amplitude change is caused by changes in thickness and not by changes in impedance of the thin layer or of the material above and below it. As the bed becomes thinner, the amplitude will eventually decrease so far that it is invisible. The thickness where this will happen is not easy to predict, because it depends on the level of seismic

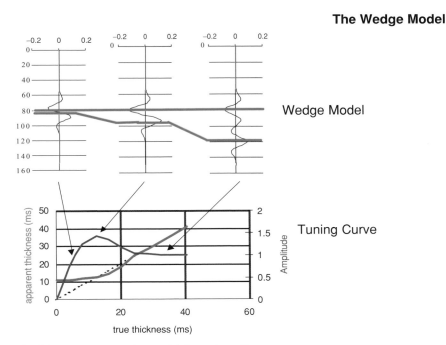

Fig. 4.2   Schematic wedge model for tuning effects.

noise, and the extent of complications from the presence of other adjacent layers. If it is, say, half the resolution limit, this implies that the standard seismic method will see only those layers whose thickness is greater than say 20 ft.

A model of a tuned hydrocarbon sand response is shown in fig. 4.2. This shows the amplitude response and apparent thickness of a sand bed encased in shale, using a zero-phase wavelet, and increasing the bed thickness from zero through the tuning range. As expected, there is a linear increase in amplitude with true thickness when the bed is thin, while the apparent thickness remains constant. There is a maximum amplitude produced by constructive interference, where the precursor of the reflection from the base of the sand is added to the main lobe of the reflection from the top of the sand. Beyond this point, the top and base of the sand are observable as separate reflectors, and the amplitude falls to the value expected for an isolated top sand reflector. Figure 4.3 shows an example on an actual seismic line. The amplitude of the gas sand reflection is highest (bright yellow) on the flanks of the structure where there is tuning between the top of the gas sand and the gas–water contact, and decreases towards the crest of the structure (orange-red) where the gas column is greater; in this particular example, the column is never great enough to resolve the gas–water contact as a separate event. In map view, the result will be a doughnut-shaped amplitude anomaly, with the highest amplitudes forming a ring around the crest at the point where the tuning effect produces the highest amplitude.

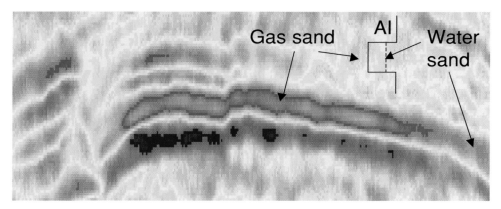

Fig. 4.3   Seismic section showing a tuned gas sand response.

Horizontal resolution of seismic data is also limited. Migration in theory collapses a diffraction hyperbola to a focus whose width will be about half the dominant wavelength (Claerbout, 1985) if data are available at all dips, or more pessimistically a width about equal to the dominant wavelength (Stolt & Benson, 1986) if the dips that can be incorporated into the migration are limited to a maximum of 30–45°. This means that the best lateral resolution we could hope for, with the 6–60 Hz wavelet, might be about 60 ft, or about 18 m. However, horizontal resolution is severely degraded by even small errors in migration velocity; as pointed out by Lansley (2000), a 0.5% error in migration velocity can degrade horizontal resolution by a factor of more than 5. In practice, therefore, horizontal resolution will often be in the range 50–100 m. This is something that should be borne in mind when making decisions on exactly where on a seismic survey a well should be located. If it is very close to a major fault there will be a risk of reaching the target on the wrong side of the fault. The horizontal resolution also limits the geological detail that we can hope to see on seismic sections, though it is sometimes possible to see more than these formulae suggest (Goulty, 1997).

4.2   Seismic stratigraphy

To understand the distribution of a reservoir sand in the subsurface, we need to see it as part of a depositional system. There are eight primary clastic (sand and mud) depositional systems (Galloway, 1998): alluvial fan, fluvial, delta, shore-zone, shelf, slope, aeolian and lacustrine. Over time, the depositional systems within a basin change; abandoned systems are buried and eventually may become reservoir rocks. Using a combination of seismic data and well control, it may be possible to understand the depositional system well enough to be able to predict sand distribution and quality in areas that have not yet been drilled. This is achieved in part by recognition of individual depositional units from the seismic data, and in part by placing them within an overall context. The latter task is aided by the concepts of sequence stratigraphy, which uses unconformity surfaces to define boundaries of packages of rocks that are of similar age and deposited within a related family of depositional systems.

Galloway (1998) gives several examples of the way in which it may be possible to infer sand distribution from detailed seismic mapping. For example, channel systems are often seen in submarine fan, fluvial, and deltaic environments. Channels scour their beds and banks during periods of high-volume flow, and deposit sediment within and around the channel during periods of lower flow. Channels may be straight to highly sinuous in plan view, broad to narrow, and shallow to deep. They may be largely erosional (depositing little sediment beyond their banks) or depositional, building large levees. In general, muddy systems tend to have narrow, deep sinuous channels with prominent levees; in such a system, sands are often narrow isolated lenticular bodies. Sand-rich systems, on the other hand, tend to have broad, low-sinuosity channels that do not have well-developed levees. Another type of sand deposition is the result of unconfined fluid flow. This is most obviously found in a marine shelf or aeolian setting, but also occurs within other environments, such as crevasse splays along rivers and turbidite lobes in submarine fans.

Within a depositional system, sandy reservoir and muddy seal associations show predictable patterns. For example, in a fluvial system the best sands are found as channel-fill deposits (e.g. point bars). Crevasse splays along the channel banks may contain sands deposited in small branching flood channels that are poorly connected to the sands of the channel fill. Muddy deposits in abandoned channels may segment the top of the channel-fill sand with shale plugs; levees on opposite sides of a channel may not be in pressure communication if the channel is mud-filled. Vertical as well as lateral facies changes may be predicted. For instance, in map view a delta consists of the delta plain with a network of distributaries, the delta front with beaches, tidal flats and channel mouths, and the submarine delta shoreface and muddy prodelta. As the delta builds out across the shelf, a corresponding vertical succession is formed: a basal muddy prodelta facies is overlain by delta front sands, capped by lenticular distributary channel fill units and sealed by mudstones deposited when delta lobes are abandoned and transgressed by the sea. The distinctive contribution of 3-D seismic is that mapping of these individual units will be much more reliable than can be achieved with a 2-D grid, and so inferences based on the shape of the bodies (e.g. channel sinuosity) will be much more reliable.

The overall depositional setting can be elucidated using the concepts of sequence stratigraphy. A useful summary of current thinking on this topic has been given by Reading & Levell (1996). As originally published, there was considerable emphasis on cycles of sea-level change as the cause of sequence development and the main control on stratigraphic facies; charts were published purportedly showing global sea-level behaviour over geological time (Vail *et al.*, 1977). The concept of a universally valid, global, sea-level curve has been questioned by many authors (see, for example, Underhill, 1991), who point out the importance of local tectonics to local relative sea-level change. However this may be, the general concept of the sequence stratigraphic model has proved useful in predicting the lithological succession at a continental shelf margin during a single cycle of relative sea-level change. Depositional systems may be described in terms of systems tracts, containing contemporaneous depositional systems that pass laterally from fluvial to deltaic to deep-water systems.

These systems tracts are often interpreted in terms of their position in the sea-level cycle, consisting of a major sea-level fall, a lowstand, a sea-level rise and a highstand. Sea-level falls result in the formation of unconformities that form the sequence boundaries, and exhibit sub-aerial exposure and a downward shift in coastal onlap. In the initial sea-level fall, there is erosion of the coastal system and deposition is confined to basin floor fans. During formation of the lowstand systems tract, a lowstand wedge of sediment is deposited that consists of leveed channel complexes of slope fans and shelf-edge deltaic complexes. As sea-level rises, a transgressive systems tract is formed; deposition is reduced in the basin and transgressive systems form on the shelf and the coastal plain. At the top of the system there is a maximum flooding surface, and the highstand systems tract is marked by systems that aggrade and eventually prograde seaward as accommodation space created by the sea-level rise decreases. Sequence boundaries can be recognised on seismic from the onlap patterns, but can be difficult to pick. Study of isolated 2-D seismic sections may miss significant features because they ignore the map view and do not see lateral changes parallel to the coast.

Each systems tract presents its own reservoir associations. Thus, in the lowstand tract, most sediment bypasses the fluvial and delta plain environments; shelf-margin delta lobes will offer reservoir targets in the delta front, and sand bodies associated with distributary channels. In the transgressive tract, stratigraphic traps may be formed in strike-parallel sand bodies such as wave/tide reworked shelf sand bars and barrier islands. The highstand systems tract often contains fluvial channel-fill sands encased in overbank mudstones.

One of the key elements of sequence stratigraphy as formulated by Vail *et al.* in AAPG Memoir 26 was the assertion that seismic reflections generally follow chronostratigraphic surfaces. Although this seems to be correct in many cases, it is not immediately obvious how to relate this to the way that seismic reflections are caused by impedance contrasts across layer boundaries, which is fundamental to the type of detailed prediction of reservoir properties discussed in chapter 5. The importance of resonance between the seismic pulse and cyclic sea-level change has been emphasised by Anstey & O'Doherty (2002). For typical sedimentation rates, a cyclic sea-level variation with a period of 1–5 million years would give rise to cyclic sedimentation patterns with thicknesses in the range 15–300 m; for a typical seismic velocity of 3000 m/s, they would have a TWT thickness of 10–200 ms, which is about the same as the period range for typical seismic waves (frequencies 5–100 Hz). The effect of the cyclic change in sea-level is to cause a change in accommodation space that will cause deposition at any particular point to change from more landward to more seaward, and back again. If there is a systematic impedance difference between landward deposits and seaward deposits (caused perhaps by their being sand-prone and shale-prone, respectively), then there will be cyclic changes in impedance. During one half-cycle the impedance will generally be increasing, so reflection coefficients at individual thin beds will all be positive, and over the other half-cycle they will all be negative. Superposed on these cycles there may be quite large impedance jumps (and resulting reflection coefficients) at depositional hiatuses. However, the cyclic effects will dominate if there is resonance with the periods found in the seismic pulse, which is why seismic reflectors tend to be chronostratigraphic markers linked to cyclic sea-level change. Incidentally, this is an additional reason why well synthetics may not match actual seismic data; the resonant reinforcement of the signal may be quite sensitive to details of the wavelet used to create the synthetic. If it does not have a smooth broad-band spectrum, then the resonant effects may be suppressed at frequencies where the wavelet spectrum is deficient.

4.3   Interpretation tools

Geological interpretation of seismic data may be simple if there is adequate well control, but in many cases the interpreter has to make inferences from the appearance of observed bodies. This may include both their external form and, if resolution is good enough for it to be visible, the geometry of internal reflections. Some ways of looking for distinctive features are as follows.

(1) Vertical sections. Standard displays as discussed in chapter 3 may be adequate to show the geometry of individual bodies, particularly if they are thick enough to show distinctive internal reflections, such as the dipping foreset beds of a delta front. The top and base of the unit containing the foresets can be picked by the same methods as used for structural interpretation, and in some cases it may be possible to map a number of vertically stacked or laterally equivalent units. A tool that may be useful is the instantaneous phase display, which is derived from the seismic trace as follows (Barnes, 1998; Taner *et al.*, 1979). Suppose we define the envelope of the seismic trace at any particular TWT as the maximum value that the trace can have when modified by applying a single phase rotation to the entire trace. In principle, this could be found by observing how the trace changes when the phase is rotated through the range 0 to 360°; at any TWT, the maximum value that the trace assumes during the rotation is the envelope or instantaneous amplitude, and the phase rotation that gives rise to this maximum amplitude is the instantaneous phase (reversed in sign). Finding these values for all times on the trace gives us the envelope and instantaneous phase traces. The actual calculation is in practice

computationally straightforward enough for many workstations to have the ability to create instantaneous amplitude and phase displays for any section in not much more than the time taken to display the data. Essentially, the instantaneous phase display looks like a seismic display with a very short gate AGC applied; amplitude information is suppressed, and peaks can be followed across the section with a constant phase of $0°$, troughs with a constant phase of $180°$, and zero-crossings with a phase of $\pm 90°$. When displayed with a suitable colour-bar (i.e. one that starts and ends with the same colour so that $+180°$ has the same colour as $-180°$), then the instantaneous phase section makes it easier for the interpreter to spot angular relationships (onlap, downlap, etc.) in low-amplitude parts of the seismic section. An example is shown in fig. 4.4. Arrows highlight places where the instantaneous phase section shows angularities more clearly than the standard reflectivity data.

(2) Horizontal sections. Time slices can reveal map-view geometry, such as channel systems. However, if there is structural dip present, the horizontal slice does not show data referring to a single stratigraphic level.

(3) Horizon slices. By slicing through the data parallel to a particular event it is possible to see a map of amplitude changes at a single stratigraphic level. This is often the best way to see channel and fan features, which are recognised by their geometry in map view. The reference horizon is usually chosen to be the strongest and most continuous marker within the sequence, as this can be autotracked most easily. This is a good way to look at the internal geometry of thin layers at or below the limit of seismic resolution; all the information is encoded in the lateral amplitude variation of the reflector. In such a case, it may be worthwhile to invert the data by the methods discussed in chapter 6, with the aim of increasing the bandwidth and thus getting slightly more information out of horizon slices through the inverted volume. With thicker layers, it can be more informative to look at amplitudes (e.g. rms average to avoid mutual cancellation of positive and negative values) within a window whose thickness is chosen so as to enclose the layer of interest; reconnaissance of the feature using vertical sections will show what window size to use. A refinement of this idea is to use 'stratal slices' (Zeng *et al.*, 2001). In this method, displays are produced of a seismic attribute (e.g. amplitude) on a geological time surface. This surface is created by linear interpolation between picked surfaces that are believed to be time-parallel reference events, e.g. marine flooding surfaces; as we saw earlier in the chapter, such sequence boundary reflections are often strong, easily picked and laterally continuous.

(4) Coherence slices. Use of horizontal slices through the coherence cube to map faults was explained in chapter 3. Horizon-parallel slices can be used to reveal map-view information about internal structure of a layer, in exactly the same way as for reflectivity. Subtle internal discontinuities can be revealed. To understand features seen in map view, they may need to be compared with their expression on vertical sections; standard reflectivity sections should be used for this purpose

Reflectivity

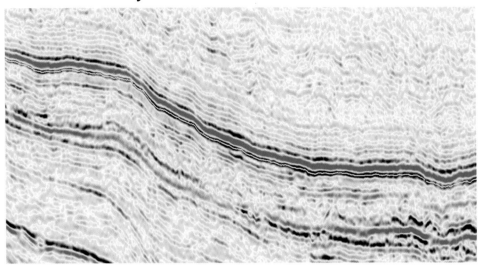

Instantaneous phase

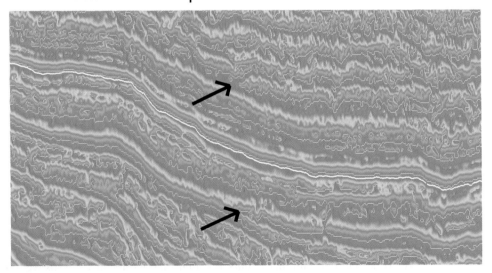

Fig. 4.4 Reflectivity and instantaneous phase displays of a seismic section.

as the method of calculating the coherence cube makes vertical sections through it hard to understand.

(5) 3-D views. The types of view discussed in chapter 7 can be very helpful in under-standing a geological feature. If the feature is marked by high-amplitude events that can be voxel-picked, it is easy to obtain a rapid first impression of its 3-D shape. Where the seismic expression of the body is less obvious, it may be necessary to

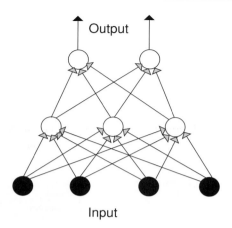

Fig. 4.5   Schematic neural network (after Gurney, 1997).

pick a top and base surface of the body by the usual manual methods, and then display the seismic between these surfaces as a semi-transparent 3-D body.

(6) An entirely different approach is to use a classification algorithm to map areas of similar seismic character. One approach to this is to use neural net software. A general introduction to the principles of neural networks may be found in Gurney (1997), for example. A simple example of such a network is shown in fig. 4.5. Input data are fed to nodes in the first layer of the network. Each arrowed path has associated with it a weight, and the input values are multiplied by the weight corresponding to the particular path that they travel. On arrival at a node, the weighted inputs are summed; if the sum exceeds a threshold value, the node sends out a high signal value (conventionally '1') to the next layer of nodes, or if the threshold value is not reached the node sends out a zero signal value. The same weighting procedure is carried out along the paths to the second layer, and the nodes in that layer sum the inputs and output a 1 or a 0 depending on whether the summed input exceeds the threshold or not. This behaviour of the nodes mimics that of nerve cells (neurons) in biological brains. The weights on the interconnecting paths determine how the system behaves. They can be determined by a learning process, in which input data are presented for which the correct output is known. For example, if we wanted to predict whether a certain layer is sand or shale from seismic data, then the input could be a set of seismic attribute values (trace amplitudes, loop widths, etc.) at a well location where it was known whether the layer was sand or shale. If we had a number of wells, some with sand and some with shale, each with its own seismic attribute values from an adjacent trace, it would be possible to adjust the neural net weights iteratively so that the output is a sand/shale flag when the seismic attribute values are supplied as input. An extension of this idea is to predict values of reservoir properties such as porosity. One way of doing this is to classify trace data according to their similarity to synthetics produced from wells with known

rock properties (Trappe & Hellmich, 2000). A different way of training a net is to ask it to cluster input data, without specifying in advance what the clusters should be; the net looks for similarities in pattern of the input data and groups the input into classes accordingly. For example, the input data to a net might be seismic trace data, over a window hung off a reference picked horizon and large enough to encompass several loops. The net then clusters the data, looking at the shape of the traces rather than their amplitude. A map is produced, coloured according to the cluster that a trace belongs to. An example is shown in fig. 4.6. Here the data have been split into 12 classes, with characteristic trace shapes as shown in the lower part of the figure. All these traces begin at the maximum of a trough, because the window used for data selection was hung off an autotracked trough. The map shows a prominent lineation (arrowed) to the north-east of which the traces are quite different from those elsewhere in the map. This lineation is inferred from well control to be the edge of a major Tertiary fan system; within the fan, to the south-west of the lineation, there are additional variations which are not well understood for lack of well calibration.

4.4   Some examples

In this section we describe examples of some of the techniques described in the previous section. Stratal slicing is demonstrated by fig. 4.7 (Zeng *et al.*, 2001). This comes from the Miocene–Pliocene section of offshore Louisiana. From well data, it is known that in this area sands are acoustically softer than shale. The polarity convention for these slices is that red = soft, so in general we expect red to correspond to sand and blue to shale. Slices at different levels show various features. In (a), we see moderately sinuous, channel-like features. Based on comparison with well penetrations, these are thought to be fluvial channels in a coastal plain environment, with fining-upward channel fill. A deeper stratal slice, (b), shows a very different channel type, with low sinuosity. Well penetrations show blocky log patterns. These are incised valley fills that contain deposits of lowstand and transgressive systems tracts. A deeper slice still, (c), shows soft red amplitudes with lobate to digitate plan-view geometry, grading into low- to variable-amplitude lobes. Wells in the channels (e.g. log 1) show an upward-fining distributary channel overlying upward-coarsening, prograding delta deposits. Delta-front deposits (log 2) contain thin interbedded sands and shales, while prodelta wells encounter shales (log 3). The overall system is interpreted as a highstand shelf delta.

Another example is shown in fig. 4.8, this time from offshore Egypt (Wescott & Boucher, 2000). These are submarine delta-front channel complexes, formed during a late Miocene–earliest Pliocene transgression, and are well imaged on horizon slices through a coherency volume. The deeper Rosetta channel complex is well defined because it is incised into underlying anhydrite; it is characterised by sharp channel edges and low sinuosity. These channels are interpreted to be sediment bypass conduits,

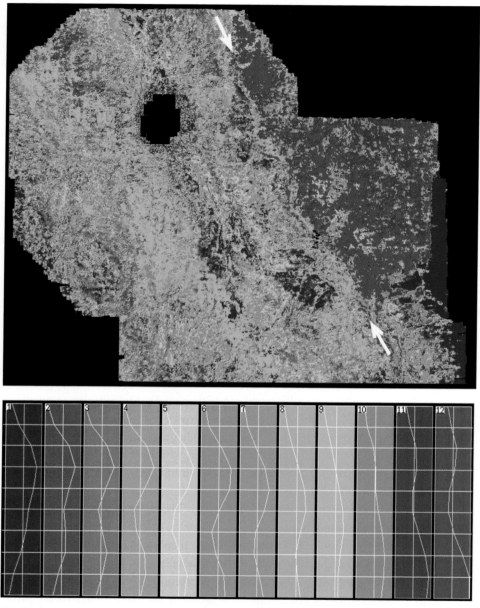

Fig. 4.6 Seismic facies classification map, based on the trace shapes in the lower part of the figure. Image produced using Stratimagic™ software (Paradigm Geophysical), which incorporates SISMAGE™ technologies developed by TotalFinaElf.

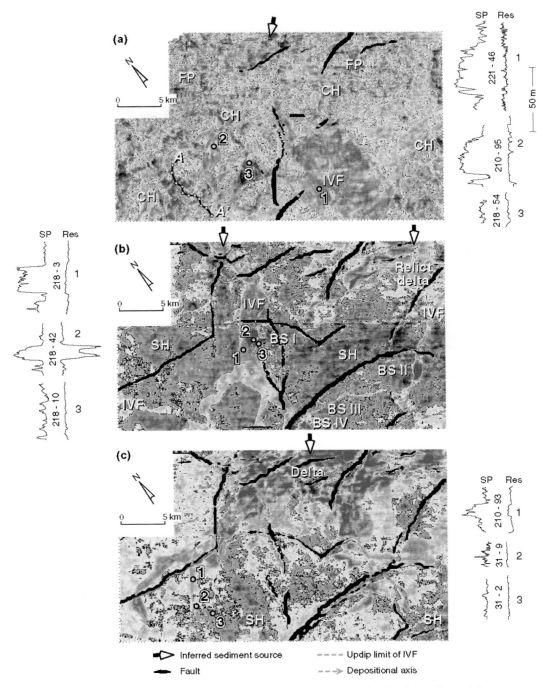

Fig. 4.7 Amplitude stratal slices showing (a) a Pliocene coastal plain, (b) an Upper Miocene incised valley fill, and (c) an Upper Miocene highstand shelf delta system. CH = channel, FP = floodplain, IVF = incised valley fill, SH = shelf. Reproduced with permission from Zeng *et al.* (2001).

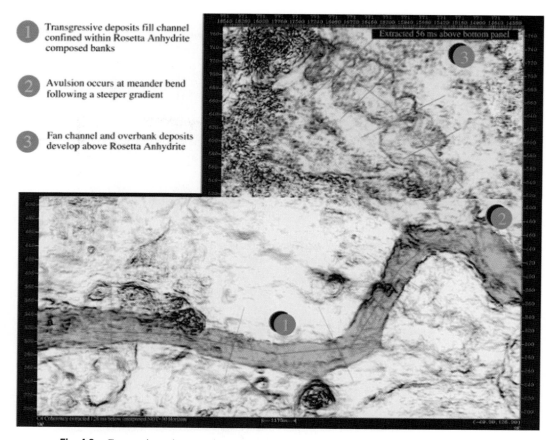

Fig. 4.8 Composite coherency image showing the Rosetta channel complex (orange), with the image of the Abu Madi channel complex (56 ms above the Rosetta; yellow) superimposed on the upper right. Light yellow outlines levee/overbank deposits associated with this channel system. Reproduced with permission from Wescott & Boucher (2000).

transporting sediment across the shelf into deeper water. They filled with sediment as transgression continued, and the system was no longer confined by the resistant evaporite canyon walls. Eventually the channel avulsed at a meander bend, following a steeper gradient, and formed the Abu Madi channel system, with higher-sinuosity channels confined within fan/overbank deposits.

An example of interpretation of downlap geometries is shown in fig. 4.9. This is from the San Jorge Basin, Argentina (Wood *et al.*, 2000). This is a fluvial system; the successive south-to-north downlap onto the M7 unconformity is attributed to lateral accretion of migrating point bars. The overlying deposits show accretionary processes from north to south, with downlap onto an unconformity separating this unit from the deeper point bars. The top of this interval is a widespread flooding surface, marked by alluvial deposits. Mapping amplitudes, on both standard reflectivity data and coherency volumes, in slices parallel to this flooding surface, allowed channel systems to be mapped and their evolution followed over time.

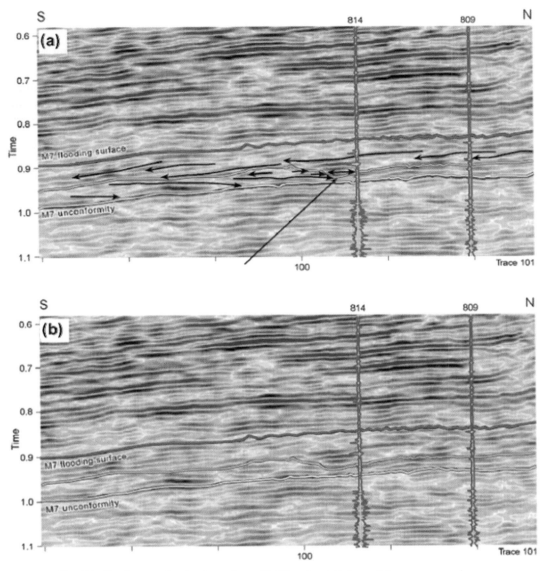

Fig. 4.9 South to north seismic section with SP and resistivity well logs superposed: (a) interpreted and (b) uninterpreted. Note incised channel south of well 814. Reproduced with permission from Wood *et al.* (2000).

4.5   Faults

As we saw in chapter 3, fault systems can be mapped in some detail from 3-D seismic, although there is room for confusion between small faults and seismic noise. It is often necessary to determine whether a mapped fault will form an adequate lateral seal to a mapped undrilled structure, or to evaluate whether a fault will be a significant barrier to fluid flow during production from a reservoir. If we can assume that faults are not in themselves hydrocarbon seals, then hydrocarbons can flow across the fault wherever permeable layers are juxtaposed across it. This can be analysed by drawing sections along the fault plane, showing the layers intersecting the fault on both the upthrown and the downthrown side (Allan, 1989). From this juxtaposition diagram, the spillpoint of a structure can be determined as the shallowest depth at which hydrocarbon migration across the fault is possible. This procedure is not as simple as it sounds. Many faults are not simple single discontinuities, but are complex zones consisting of a series of interconnected fault segments (Knipe *et al.*, 1998). Often, a large single fault would be judged an effective barrier to fluid flow from the juxtaposition diagrams, but the equivalent ensemble of small-displacement faults might not be. Very careful mapping of the faults is then required, using amplitude, coherency and dip maps, together with review of vertical sections, to establish the fault pattern. This is particularly difficult to do for the small faults that grade into seismic noise.

A complication is that the fault plane itself may be an effective hydrocarbon seal, even though permeable strata are juxtaposed across it. This can be the result of smearing of clay along the fault plane during displacement along the fault. Various methods for predicting the presence of clay smear have been summarised by Foxford *et al.* (1998). These may be based on:

(i) the percentage of shale or mudstone layers in the faulted sequence,
(ii) the percentage of shale in the sequence that was moved past any point on the fault surface,
(iii) the along-fault distance in the slip direction of a point on the fault surface from a potential shale source layer, and the thickness of the layer.

Foxford *et al.* used the second of these approaches (the shale gouge ratio, SGR), and found that an SGR of less than 20% was characteristic of fault zones that did not contain shale gouge in their particular study. Similar cutoff values have been found in other studies. As the SGR increases above this level, the fault plane becomes a more effective seal, able to hold a longer hydrocarbon column over geological time or sustain a larger pressure drop across it on a field production timescale. However, the thickness of the shaley gouge can be highly variable and unpredictable. It is therefore difficult to use SGR in a quantitative way to determine fault permeabilities (Manzocchi *et al.*, 1999). Compilation of data from existing fields is needed to reduce these uncertainties (Hesthammer & Fossen, 2000).

References

Allan, U. S. (1989). Model for hydrocarbon migration and entrapment. *American Association of Petroleum Geologists Bulletin*, **73**, 803–11.

Anstey, N. A. & O'Doherty, R. F. (2002). Cycles, layers, and reflections. *The Leading Edge*, **21**, 44–51.

Barnes, A. (1998). The complex seismic trace made simple. *The Leading Edge*, **17**, 473–6.

Claerbout, J. F. (1985). *Imaging the Earth's Interior*. Blackwell Scientific Publications, Oxford.

Foxford, K. A., Walsh, J. J., Watterson, J., Garden, I. R., Guscott, S. C. & Burley, S. D. (1998). Structure and content of the Moab Fault Zone, Utah, USA, and its implications for fault seal prediction. In: *Faulting, Fault Sealing and Fluid Flow in Hydrocarbon Reservoirs* (eds. G. Jones, Q. Fisher & R. J. Knipe), *Spec. Publ. Geol. Soc. Lond.*, **147**.

Galloway, W. E. (1998). Clastic depositional systems and sequences: applications to reservoir prediction, delineation, and characterization. *The Leading Edge*, **17**, 173–9.

Goulty, N. R. (1997). Lateral resolution of 2D seismic illustrated by a real data example. *First Break*, **15**, 77–80 (and correction in *First Break*, **15**, 331–2).

Gurney, K. (1997). *An Introduction to Neural Networks*. UCL Press, London.

Hesthammer, J. & Fossen, H. (2000). Uncertainties associated with fault sealing analysis. *Petroleum Geoscience*, **6**, 37–45.

Knipe, R. J., Jones, G & Fisher, Q. J. (1998). Faulting, fault sealing and fluid flow in hydrocarbon reservoirs: an introduction. In: *Faulting, Fault Sealing and Fluid Flow in Hydrocarbon Reservoirs* (eds. G. Jones, Q. Fisher & R. J. Knipe), *Spec. Publ. Geol. Soc. Lond.*, **147**.

Lansley, M. (2000). 3-D seismic survey design: a solution. *First Break*, **18**, 162–6.

Manzocchi, T., Ringrose, P. S. & Underhill, J. R. (1999). Flow through fault systems in high porosity sandstones. In: *Structural Geology in Reservoir Characterization* (eds. M. P. Coward, H. Johnson & T. S. Daltaban), *Spec. Publ. Geol. Soc. Lond.*, **127**.

Reading, H. G. & Levell, B. K. (1996). Controls on the sedimentary rock record. In: *Sedimentary Environments: Processes, Facies and Stratigraphy* (ed. H. G. Reading). Blackwell Science, Oxford.

Stolt, R. H. & Benson, A. F. (1986). *Seismic Migration*. Geophysical Press.

Taner, M. T., Koehler, F. & Sheriff, R. E. (1979). Complex seismic trace analysis. *Geophysics*, **44**, 1041–63.

Trappe, H. & Hellmich, C. (2000). Using neural networks to predict porosity thickness from 3D seismic data. *First Break*, **18**, 377–84.

Underhill, J. R. (1991). Controls on Late Jurassic seismic sequences, Inner Moray Firth, UK North Sea: a critical test of a key segment of Exxon's original global cycle chart. *Basin Res.*, **3**, 79–98.

Vail, P. R., Mitchum, R. M. & Thompson, S. (1977). Seismic stratigraphy and global changes of sea level. Part 3: relative changes of sea level from coastal onlap. In: *Seismic Stratigraphy – Applications to Hydrocarbon Exploration* (ed. C. E. Payton), *Mem. Am. Ass. Petrol. Geol.*, **26**, Tulsa.

Wescott, W. A. & Boucher, P. J. (2000). Imaging submarine channels in the western Nile Delta and interpreting their paleohydraulic characteristics from 3-D seismic. *The Leading Edge*, **19**, 580–91.

Widess, M. B. (1973). How thin is a thin bed? *Geophysics*, **38**, 1176–80.

Wood, L. J., Pecuch, D., Schulein, B. & Helton, M. (2000). Seismic attribute and sequence stratigraphic integration methods for resolving reservoir geometry in San Jorge Basin, Argentina. *The Leading Edge*, **19**, 952–62.

Zeng, H., Tucker, F. H. & Wood, L. J. (2001). Stratal slicing of Miocene–Pliocene sediments in Vermilion Block 50 – Tiger Shoal Area, offshore Louisiana. *The Leading Edge*, **20**, 408–18.

5 Interpreting seismic amplitudes

In areas with favourable rock properties it is possible to detect hydrocarbons directly by using standard 3-D seismic data. Amplitude interpretation is then very effective in reducing risk when selecting exploration and production drilling locations. Not all areas have such favourable rock physics, but it is always useful to understand what seismic amplitudes may be telling us about hydrocarbon presence or reservoir quality. As well as amplitudes on migrated stacked data, it is often useful to look at pre-stack data and the way that amplitude varies with source–receiver offset (AVO). The first step is to use well log data to predict how seismic response will change with different reservoir fluid fill (gas or oil or brine), with changing reservoir porosity, and with changing reservoir thickness. Then we can use this understanding to interpret observed changes in seismic amplitude or other measures of the size and shape of individual seismic loops. In principle this process can also be used to interpret amplitudes on 2-D seismic data, but as we saw in chapter 1 the power of 3-D seismic lies in the ability to make maps based on very densely sampled data, allowing us to see systematic amplitude changes that are only just above the noise level.

5.1 Basic rock properties

We shall consider in detail only *isotropic* rocks, where the seismic velocities are independent of the direction of propagation through the rock. In practice, many rocks are *anisotropic*. The velocities of horizontal and vertical paths may be different, perhaps owing to fine-scale internal layering. Horizontal velocities may vary with azimuth, perhaps owing to cracks aligned in a particular direction. These effects are complicated and it is difficult to obtain the rock parameters needed to model them. The isotropic case is much simpler: two rock properties control the response to sound waves. These are the acoustic impedance and the ratio of compressional to shear wave velocity (V_p/V_s). The acoustic impedance is simply the product of compressional velocity and density:

$$AI = V_p\rho.$$

Compressional waves (P-waves) differ from shear (S) waves in the direction of particle motion as the wave propagates through the rock. For P-waves the motion is parallel to the direction of travel of the wave, whereas for S-waves it is perpendicular. The P- and S-wave velocities are related to different rock properties. When P-waves propagate through a rock, there are changes in the volume of individual particles, whereas S-wave propagation causes bending without change of volume. Standard seismic sources emit P-waves almost entirely, so we usually see S-waves directly only when P-waves have been partly converted to S-waves on reflection at an interface. Standard seismic processing concentrates on using P-waves to form an image of the subsurface. However, as we shall see, the shear properties of the rock are important to understanding AVO.

Sometimes a quantity called Poisson's ratio (σ) is used instead of the V_p/V_s ratio. It is given by

$$\sigma = \frac{0.5 - \left(\frac{V_s}{V_p}\right)^2}{1 - \left(\frac{V_s}{V_p}\right)^2}.$$

Figures 5.1 and 5.2 show typical values of these parameters for some common rock types.

5.2 Offset reflectivity

We saw in section 3.1.1 how the reflection coefficient at an interface depends on the acoustic impedance contrast across it for the case of normal incidence. In the real world, seismic data are always acquired with a finite separation between the source and receiver (usually termed the *offset*). This means that reflection will be much more complicated, because part of the P-wave energy will be converted into a reflected and transmitted shear wave. The equations describing how the amplitudes of the reflected and transmitted P- and S-waves depend on the angle of incidence and the properties of the media above and below the interface were published by Zoeppritz (1919); the amplitudes depend on the contrast in Poisson's ratio across the interface, as well as the acoustic impedance change. Figure 5.3 shows an example of how the amplitudes depend on incidence angle for a particular interface.

For a plane interface the relationship between the P-wave angles of incidence and transmission is given by Snell's Law (fig. 5.4):

$$\frac{\sin \theta_2}{V_2} = \frac{\sin \theta_1}{V_1}.$$

If velocity increases with depth across the interface, then there will be an incidence angle for which the transmission angle is 90°. This is the critical angle, at and beyond which there is no transmitted P-wave, and therefore a high reflection amplitude.

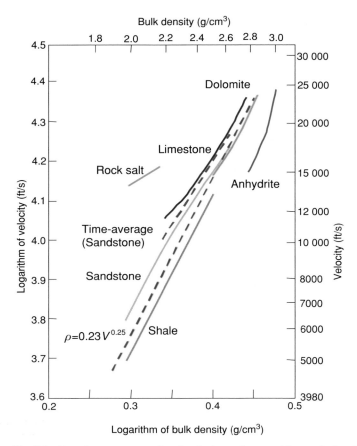

Fig. 5.1 Density vs compressional velocity, redrawn with permission from Gardner *et al.* (1974).

The Zoeppritz equations are rather complicated. It is easy enough to write software to generate curves like those shown in fig. 5.3, but it is also helpful to have approximations that give more insight into the underlying relations between reflectivity and rock properties. Useful approximations for the PP reflection coefficient (i.e. both incident and reflected waves of P type, the most common situation) have been given by Aki & Richards (1980) and by Shuey (1985). Approximately,

$$R(\theta) = A + B \sin^2 \theta + C \sin^2 \theta \tan^2 \theta$$

where

$$A = 0.5 \left(\frac{\Delta V_p}{V_p} + \frac{\Delta \rho}{\rho} \right)$$

$$B = 0.5 \frac{\Delta V_p}{V_p} - 2 \left(\frac{V_s}{V_p} \right)^2 \left(2 \frac{\Delta V_s}{V_s} + \frac{\Delta \rho}{\rho} \right)$$

Poisson Ratio and Lithology

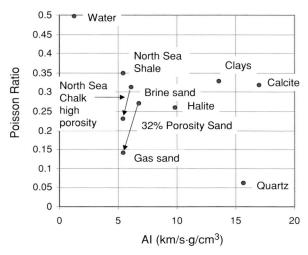

Fig. 5.2 Poisson's ratio vs acoustic impedance.

Energy Partitioning at an Interface

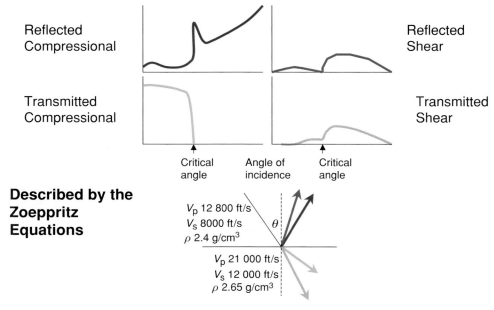

Described by the Zoeppritz Equations

Fig. 5.3 Energy partitioning at an interface, after Dobrin (1976), *Introduction to Geophysical Prospecting*, reproduced with permission of the McGraw-Hill Companies.

Snell's law

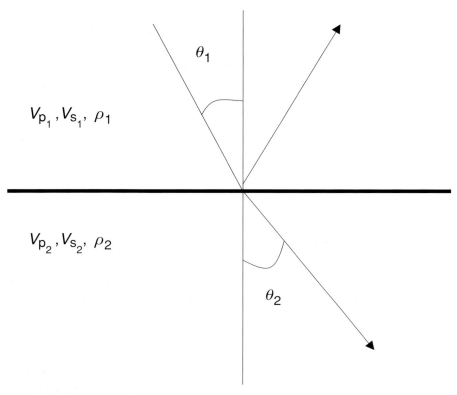

Fig. 5.4   Compressional wave reflection and transmission angles.

and

$$C = 0.5\frac{\Delta V_p}{V_p},$$

where $R(\theta)$ is the reflection coefficient at the incidence angle θ (strictly speaking, at the average of the angle of incidence and the angle of transmission as determined by Snell's Law), V_p is the average of the P-wave velocity on the two sides of the interface, V_s the average S-wave velocity, and ρ the average density, and the quantities preceded by Δ are the differences in the relevant parameter across the interface.

These equations can be used to model the seismic response when the rock properties are known. The simplest model is the single interface between two layers of different properties, and is often already very instructive. However, it is often necessary to understand the seismic response of a thin layer, for example to study how the seismic expression of a reservoir sand changes as it approaches pinchout. This can be examined using a wedge model, a single layer whose thickness is systematically varied from zero to the desired maximum. A further step toward realism is to construct a model using all the different layers within some interval, as recognised from wireline well logs. This is important for relating well data to real seismic response, but the interference effects between a whole stack of interfaces can be hard to understand unless the main individual layers are modelled separately first.

For small incidence angles, a further simplification of the Zoeppritz equations is possible; the term in the above formula involving C can be neglected. This is certainly the case out to incidence angles of $30°$ or so, and is often a reasonable approximation out to 40–$45°$, beyond which the data are in any case often muted out from gathers because of NMO stretch and the presence of direct arrivals. Then we can write the reflection coefficient in terms of the normal incidence reflectivity R_0 and the *AVO gradient*, G:

$$R(\theta) = R_0 + G \sin^2 \theta.$$

5.3   Interpreting amplitudes

Sometimes we can interpret fluid fill from amplitudes on seismic data. Before we can do so, we need to have reasonable confidence in the validity of the amplitudes in the seismic dataset. As explained in chapter 2, modern processing will try to avoid any steps that cause amplitude artefacts. Ideally, we would like to have seismic data where amplitudes are everywhere proportional to reflectivity. This is not achievable, but what can be done is to make sure that local lateral variation of amplitude (over a distance of, say, a kilometre) on a particular group of reflectors is proportional to reflectivity change. We can often assume that the average absolute reflectivity over a long time window varies little, so a long-gate AGC can be applied to the data. It is essential, though, that the gate is long enough (1000 ms or more) to avoid destroying the lateral variations we are looking for; the gate should include many reflectors, so that the target event makes very little contribution to the average amplitude in the window. Calibration of amplitude to reflectivity is possible from a well tie, but the calibration is valid only over a limited interval vertically. In any case, it is a good idea to inspect the entire section from surface to the target event and below; if amplitude anomalies at target level are seen to be correlated with overlying or underlying changes (high or low amplitudes due to lithology or gas effects, or overburden faulting, for example), then they should be treated with suspicion. Such a correlation might have a genuine geological cause, but careful thought is needed to establish that the effect is not an artefact. Following the amplitude anomaly through the seismic processing sequence from the raw gathers may be helpful; this may reveal an artefact being introduced in a particular processing step.

To recognise hydrocarbon effects (Direct Hydrocarbon Indicators, DHIs) for what they are, we need to know what to expect. The sketch in fig. 5.5 shows what to look for

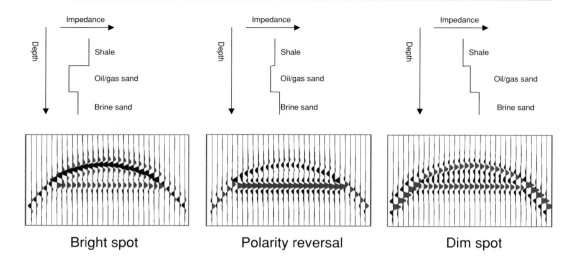

Red = hard loop (impedance increase)   Blue = soft loop (impedance decrease)

Fig. 5.5   Schematic models for oil and brine sand response.

in three possible cases. This general idea is valid for both standard full-offset stacks and the near and far trace sub-stacks that we shall discuss in the next section, although which picture applies may depend on what offset range we look at. On the left we have the case where a brine sand is softer (lower impedance) than the overlying shale. The presence of oil or gas will make it softer still. We therefore see an increase in amplitude over the crest of the structure where the hydrocarbon is present. This is the classic 'bright spot'. Of course, an amplitude increase might also be caused by a lithological effect, for example a decrease in sand impedance due to porosity increase. A key test is that we would expect the bright spot, if it is really the result of hydrocarbons, to conform to structure; in map view, the amplitude change should follow a TWT contour (or strictly speaking, a depth contour after time–depth conversion, though over a limited area the TWT and depth contours are often similar in shape). We also hope to see a 'flat spot' at the hydrocarbon–water contact. This is always a hard reflector (impedance increase). The flat spot should be at the same TWT as the amplitude change. (If we have both oil and gas present, then we may see two flat spots, one at the gas–oil contact and one at the oil–water contact, with matching amplitude steps in the top reservoir reflector, from very bright in the gas-bearing part to bright in the oil-bearing part.) In the middle, we see the case where the brine sand is hard relative to the overlying shale, but the hydrocarbon sand is soft. The top sand will be a hard loop below the fluid contact and a soft loop above it, with a polarity change at the contact. This case is often difficult to interpret with confidence, particularly if the structure is affected by minor faulting. It is often possible to keep the top sand pick on a hard loop across the crest by interpreting a small fault near the contact. Inspection of lines in different azimuths across the crest

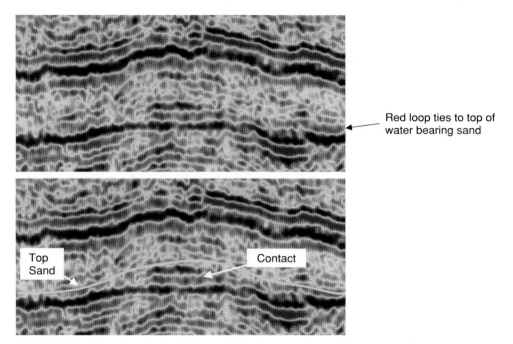

Fig. 5.6 Seismic section showing hydrocarbon indicators expected from fig. 5.5 (dim spot case).

will help the interpreter to say if there is actually a consistent polarity flip. On the right, we see the case where the brine sand is hard relative to the shale, and the hydrocarbon sand, though softer than the brine sand, is still hard relative to the shale. In this case, we expect to see a 'dim spot', an amplitude decrease at top reservoir. It should conform to structure and to the TWT of the flat spot. This is not as easy to recognise as the bright spot case.

   An example of a dim spot is shown in fig. 5.6. Brine sands are expected to be hard and oil sands very weakly hard relative to the overlying shale. We then expect the top sand reflector to dim in the oil-bearing interval, which is exactly what is seen in the real seismic (fig. 5.6). The oil–water contact (OWC) is a strong flat event. In such a case, it is very easy to make the erroneous pick of a strong continuous 'top sand' over the structure by jumping from the top brine sand to the OWC and back to the top brine sand on the other side of the structure, missing the oil accumulation altogether. Confidence in the actual interpretation shown in fig. 5.6 is improved if it passes several tests.

(1) The amplitude change from top brine to top oil sand, and the amplitude of the OWC event, are reasonably consistent with the expected values for fluid effects (modelled by methods to be discussed later). The OWC must have the correct polarity, corresponding to an increase in impedance downwards across the interface.

(2) The OWC should be flat (usually, though velocity effects can cause it to be tilted in TWT and contacts are sometimes not flat in depth. Also, interference with strong

bedding reflectors can cause the flat event to be broken up into a series of segments that may individually appear to be slightly tilted, although the ensemble remains flat).

(3) There should be apparent thickening of the isochore in the interval above the top sand if the alternative 'continuous strong loop' interpretation is used.

(4) The flat event at the OWC should run horizontally across inclined bedding, resulting in apparent reflector terminations below it.

(5) Crucially, the amplitude dimming, the flat spot extent and the apparent isochore change should be consistent in map view with each other and with a mapped trap, e.g. a dip closure. The amplitude change should follow a structural contour if it is indeed caused by a change in fluid type at the downdip edge of a trap. This is where 3-D seismic can make a big contribution. Both the amplitude map and the structural map are much more detailed than could be achieved using a grid of 2-D data, so this test is much more rigorous.

If all the tests are passed then it is possible to have a high degree of confidence in the interpretation of fluid fill. It is quite usual for the evidence not to be so clear-cut, however. In particular, a failure of amplitudes to fit structure may be caused by a stratigraphic element of the trapping mechanism, or by complications due to lateral changes in rock properties (e.g. porosity). The evidence then needs to be weighed carefully together with geological understanding. Is there enough well control for us to be reasonably confident of sand and shale properties? Is the seismic data quality adequate? For example, flat events may be multiples of sub-horizontal events higher in the section. If so, they can certainly cut across bedding, as expected for a fluid contact, but will be likely to continue across the top-seal as well as the reservoir. It is particularly suspicious if the flat event is one of a whole suite at different TWTs, which points strongly towards it being a multiple. It is often possible to see very weak flattish events on a seismic section if they are looked for hard enough, and they are often multiples that have been reduced in amplitude but not quite eliminated during processing.

In many cases, lateral amplitude changes are related to changes in porosity rather than fluid fill. This is particularly true for well-consolidated sands and carbonates. Figure 5.7 shows how the impedance of the Chalk in the North Sea is strongly affected by porosity, but relatively little by fluid fill. In this case, the Top Chalk will be an impedance increase (red trough with the usual North Sea polarity convention) if the porosity is less than 35%, changing to a decrease (blue peak) for higher porosities. At constant porosity, the difference between the average impedance trend (in blue) and the lower values with hydrocarbon fill (in dotted black) is quite small; a similar impedance change could be caused by quite a small change in porosity at constant fluid fill. Conversely, the large impedance change caused by porosity variation within the usually observed range (say 10–40%) is much greater than would be caused by any change in fluid fill. If there is enough well control to calibrate the relationship, it may be possible to infer porosity

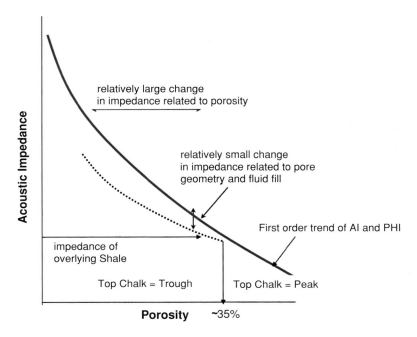

Fig. 5.7 Chalk impedance versus porosity (after Campbell & Gravdal, 1995).

values from the Top Chalk amplitude. A similar approach can be adopted to deduce sand porosity from seismic. However, this method assumes that the rock above the interface (e.g. a caprock shale) does not vary laterally. If there is a risk that it does, then the inversion methods discussed in chapter 6 are a better way to estimate reservoir porosity.

A completely different sort of DHI is the gas chimney. This occurs where gas has leaked from a deeper level into the overburden, typically along a fault plane, but the overburden is mainly shale with limited permeable zones (e.g. silts). The result is a diffuse cloud of gas-bearing material, typically with low saturations. There may be a few high-amplitude gas sand reflections at the top or within the body of the cloud, but in general scattering and absorption cause amplitudes to be much reduced below and within it, so that amplitude measurements are usually meaningless. There is often an apparent sag in TWT of reflectors below the cloud, due to the velocity decrease in the gas-bearing layers; this can cause great difficulty for accurate structural mapping in depth. Shear-wave data, which are almost unaffected by the gas, may be the best way to image the horizons below the cloud. This is often important because although the gas saturations within the chimney are too low to be of any economic value, the presence of the chimney points to the possible presence of a leaking trap below it.

Tuning is a complication for the study of amplitudes. As we saw in section 4.1, amplitudes from a thin bed can be greater or smaller than the value expected for a single interface, depending on the thickness of the bed relative to the seismic wavelength. It is often possible to observe this effect on seismic sections, manifested as an amplitude maximum at a particular apparent bed thickness. Sometimes it may be necessary to model this effect and allow for it when interpreting amplitude variation. For bed thickness less than the maximum of the tuning curve, it may be possible to map bed thickness using the linear amplitude–thickness relation, although this will work only if there is no lateral variation in acoustic impedance of the bed and the material encasing it. Accurate thickness prediction also depends on using a correct tuning curve, which in turn depends on having an accurate estimate of the wavelet present in the data.

5.4   AVO analysis

AVO response can be classified into four categories (fig. 5.8) depending on the values of R_0 and G, the normal incidence reflectivity (sometimes referred to as *intercept*) and gradient values defined in section 5.2. Figure 5.8 shows typical responses for different shale–sand interfaces (i.e. typical top reservoir in a clastic sequence); a shale–sand interface usually exhibits a negative gradient, i.e. the reflection coefficient becomes more negative with increasing offset. Class I response is characterised by an increase in impedance downwards across the interface, causing a decreasing amplitude with increasing incidence angle. Class II response has small normal incidence amplitude

Classes of AVO Response

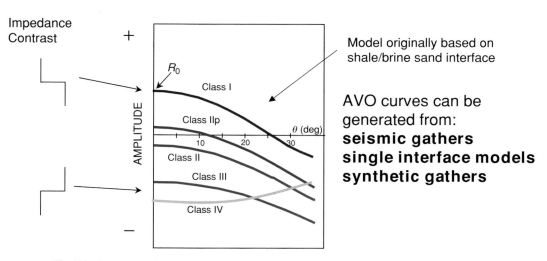

Fig. 5.8   The AVO classes. Modified after Rutherford & Williams (1989), Ross & Kinman (1995) and Castagna & Swan (1997).

Poisson Ratio, AI and Shale/Sand AVO Responses

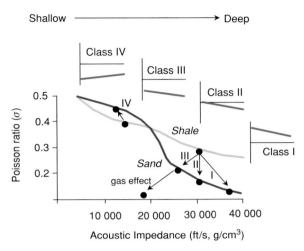

Fig. 5.9 Schematic plot of Poisson's ratio against acoustic impedance.

(positive or negative) but the AVO effect leads to high negative amplitudes at far offsets. Some authors distinguish a class IIp, where the zero incidence response is positive and therefore there is a polarity reversal at an intermediate offset. Class III responses have large negative impedance contrasts and the negative gradient leads to increasing amplitude with increasing angle. Class IV response occurs where a large negative amplitude decreases slightly with offset. Confusingly, a reflector is sometimes referred to as exhibiting *positive AVO* if the amplitude, irrespective of sign, increases with offset.

In a clastic sequence, the AVO classes are related to differences in consolidation of sands and shales with depth (fig. 5.9). The general increase in impedance with depth (fig. 5.10) reflects the decrease in porosity due to compaction. Class I responses are characteristic of deep well-consolidated sections, and class III responses of relatively unconsolidated shallow sediments. Class IV can occur in very unconsolidated sands, shallower than about 1000 m, or where soft sands are found below a non-clastic hard layer. The feature that distinguishes it from class III is the very low gradient; in practice it is often hard to say in real seismic gathers whether a low gradient is positive or negative because of scaling issues to be considered shortly.

A common method of AVO analysis is the AVO crossplot. R_0 and G can be calculated for each loop on every CMP gather in a seismic survey, by measuring the amplitude and calculating the best fit to a plot of $R(\theta)$ against $\sin^2 \theta$. The resulting pairs of R_0, G values can be charted on the crossplot and, as we shall see shortly, may give us information on fluid fill and lithology. Of course, we do not directly observe the incidence angle θ, but

Acoustic Impedance

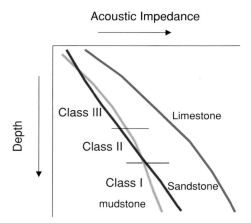

Fig. 5.10   An example of impedance trend curves, based on Gregory (1977).

instead have to calculate it from the source–receiver offset. A very rough approximation would be to assume that the ray-path from surface to reflector is straight, but this invariably underestimates the angle of incidence because of refraction effects caused by velocity increase with depth. Ray-tracing through the overburden velocity structure will give an accurate estimate, but an approximate formula for the incidence angle at offset X and zero-offset travel time T_0 is

$$\sin^2 \theta = \left(\frac{V_{\text{int}}}{V_{\text{rms}}} \right) \left(\frac{X^2}{X^2 + (V_{\text{rms}} T_0)^2} \right),$$

where V_{int} is the interval velocity at travel time T_0, and V_{rms} is the rms velocity from the surface to T_0 (i.e. approximately, the stacking velocity).

Measuring AVO gradient from real gathers is not entirely straightforward. Residual moveout can distort the measurement. Figure 5.11 shows traces from a CDP gather, where correction for NMO has not been accurate, and as a result the reflectors are not flat. The true AVO gradient would be found by measuring the amplitude along the peak or trough of a reflection. If the amplitudes are taken instead at a constant time sample, as may be the case with software designed for bulk processing of data, then the gradient will be too high. One way round this is to use trace-to-trace correlation to follow the loop across the gather (in effect, to autotrack it). This will not work in the case of a class IIp response, because of the polarity reversal. A further problem in estimating AVO gradient is that there are bound to be scaling issues between near and far offsets. A spherical divergence correction will have been applied during processing of the data to correct for the systematic decrease of amplitude with offset due to the increase in length of the travel path. If this correction is inaccurate, then there will be a tendency for all reflectors to show the same change in amplitude with offset, e.g. a systematic decrease. There are other possible causes of such systematic effects: the seismic source may emit a stronger signal in the vertical than in an oblique direction, and receiver

Residual Moveout and AVO

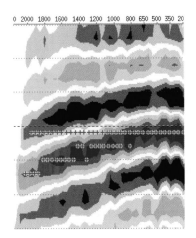

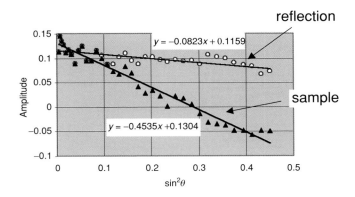

Fig. 5.11   Traces from a CDP gather (offset increasing to the left) and measured amplitudes.

sensitivity can similarly favour the vertical arrival. This is not necessarily a problem if we are interested in looking for lateral change in AVO response on a particular reflector, but it will confuse comparison of the measured response with models based on well data. It may be possible to correct the error by applying a scaling factor if we understand the cause well enough. Alternatively, we can normalise the amplitude of a target reflector against that of another reflector (or group of reflectors) of known AVO response. One way to do this is to compare the seismic trace data with well synthetics over a range of incidence angles. It is possible to calculate a well synthetic for any angle of incidence by using the Zoeppritz equations to work out the reflection and transmission coefficients at every interface, so long as we have a shear sonic log (or can predict one using the methods of section 5.5.5). One of the benefits of elastic inversion, described in chapter 6, is that it forces a careful study of such well ties to be made, to determine the wavelet amplitude at different offsets. Where there is no well control, it will be necessary to make some assumption about how amplitudes should on average behave across the offset range. This will depend on whether we expect to have an even balance of class I and class III responses, or a majority of one or the other. If we expect an even balance, then the average amplitude over a series of reflectors in a long gate can be used to scale the amplitude of the target event. In general, amplitude scaling is a major source of uncertainty. It is much easier to use AVO qualitatively, as a tool to look for lateral variation in reflector properties (e.g. to recognise pay zones), than to use it to make quantitative predictions, e.g. of reservoir porosity.

Noise on the trace data tends to have quite a large effect on the gradient calculation. A more robust approach to poor data is to use partial stacks. The simplest method is to divide the traces into two sets, nears and fars, with equal numbers of traces in each. The

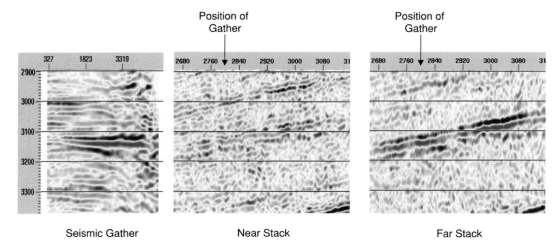

Fig. 5.12   Near and far stacks for a class III sand. Polarity: red = soft (impedance decrease downwards).

near and far data can then be stacked and migrated. The resulting near and far volumes can be loaded to an interpretation workstation and treated in exactly the same way as the full offset range data. Figure 5.12 shows an example for the case of a class III sand with marked brightening to the far offsets. The result is that the far stack shows a very strong event at the target level. This approach has several advantages.

(1) The signal to noise ratio of the near and far stacks is high compared with that of a gather.

(2) The horizon interpretation can be copied from the full to the far and near offset data; if residual moveout (due to inaccurate NMO correction) is a problem, the small adjustments to centre the horizons on the appropriate loop can be done semi-automatically. It is then possible to work with amplitude maps of a given reflector on nears and fars as a way of extracting AVO response. This removes much of the residual moveout problem encountered in gradient estimation, though the problem will still cause a degree of mis-stacking and so affect the amplitudes of the near and far displays. Sometimes, in the case of a class IIp response or where complex interference causes big changes in the appearance of reflectors from nears to fars, it may be hard to recognise corresponding reflectors on the two sub-stacks; modelling the expected response from a well dataset may be helpful.

(3) Scanning through the appropriate sub-stack volume (especially the far traces for class III) is a quick way to look for anomalous amplitudes. It is much faster than scanning through gathers; the volume of data is less, of course, but also it is possible to get an immediate impression of how amplitude anomalies relate to structure. Being able to view far and near trace sections together (in adjacent windows on the screen) is a powerful aid to understanding.

Angle Mutes from Processed Gather

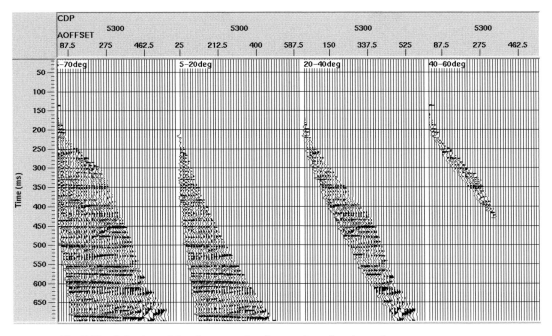

Fig. 5.13   Selection of data from a gather for creation of angle stacks.

(4) Pre-stack migration is becoming routine, and is the best approach for serious AVO study as it provides some guarantee that the traces in a gather all belong to the same subsurface reflection point. Where this has not been done, migration of the near and far sub-stacks is a cheap way of positioning the stacked amplitude data in the right place in space.

A possible problem with near and far offset stacks is that the sub-stacks represent different incidence angle ranges at different two-way times. This can make it more difficult than it need be to compare observed data with a modelled response for a particular incidence angle. The solution is to produce *angle stacks*; these are made by stacking data within an angle range calculated by ray-tracing or the formula given above. Figure 5.13 illustrates the principle.

The presence of different fluid fills (brine, oil or gas) affects the R_0 and G values of a reflector. In general, the progression from brine to oil to gas will move R_0 and G for the reflector at the top of the reservoir towards more negative values. Some schematic examples are shown in fig. 5.14. Sand A has small positive R_0 and negative G for the brine and oil cases and negative R_0 for gas. The response will be class II for brine or oil, and class III for gas. The expected stacked reflection from top sand is also shown (red = impedance increase downwards = positive reflection coefficient). For brine

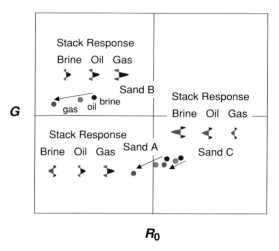

Fig. 5.14   Schematic sand responses on the R_0–G crossplot.

sand this is a red loop; the reflection coefficient decreases with increasing offset, but remains positive so that the summed response over all offsets is positive. The oil sand starts off with a small positive reflection coefficient at zero-offset, and reverses sign as it becomes more negative with increasing offset. Summing over all offsets results in partial cancellation of the positive and negative amplitudes, leaving a weak blue loop. The gas case reflection coefficient starts off negative and becomes more so with increasing offset, so the summed response is a strong blue loop. Sand B has negative R_0 with positive G, i.e. a class IV AVO response. The stacked top sand amplitude is a soft loop that increases in amplitude from brine to oil to gas, owing to the change in R_0. Sand C has a positive R_0 and negative gradient, so exhibits class I behaviour. The top sand reflector is a red loop that decreases in amplitude from brine to oil to gas, because of the decrease in R_0.

A crossplot for a top sand reflector is shown in fig. 5.15(a). The brine and oil sand cases form separate trends. Within each trend, it is porosity (increasing to the left) that causes the points to be strung out along the trend line. The oil trend is systematically displaced away from the brine trend, towards more negative values of R_0 and G. If there was reasonable seismic coverage (as is likely with 3-D data), then it would be possible to fit a trend line to the brine sand and use distance from this trend as a predictor for pay sand (fig. 5.15(b)). In this example, the points are colour-coded according to TWT, which allows us to see that the points furthest toward the bottom left (green ellipse) are slightly shallower than the points (red ellipse) nearer to the main trend. In this case, the data come from a drilled anticlinal structure that contains both gas and oil. The points in the green ellipse are from the top of the gas reservoir, the points in the red ellipse from the top of the oil reservoir. Projecting data onto the perpendicular to the trend line, as shown, is equivalent to making a weighted sum of intercept and gradient. From

(a)

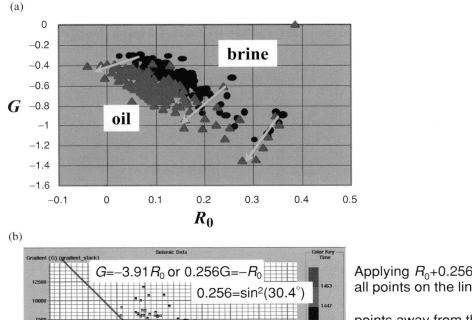

(b)

Applying $R_0 + 0.256G$ makes all points on the line zero

points away from the line will be −ve or +ve

This can be viewed as a projection

Fig. 5.15   (a) and (b) Examples of intercept–gradient crossplots for top sand in brine and oil cases.

Shuey's equation, we see that this is equivalent to creating a stack at some particular angle of incidence. An angle stack for a small range around this value would therefore be the best way of looking for pay sand. Various 'fluid factor' discriminants have been proposed along these lines, but in practice it is better to make a crossplot of the real seismic data and choose the best angle stack by the calculation shown in fig. 5.15(b). Note that in this example, the slope of the trend line is quite high at about −4. This means that an angle stack at around 30° would be the best fluid discriminant. Modelling the expected trend based on well data, considering the effects of varying porosity and

AVO Mapping -
Weighted Stack or Equivalent Angle Maps

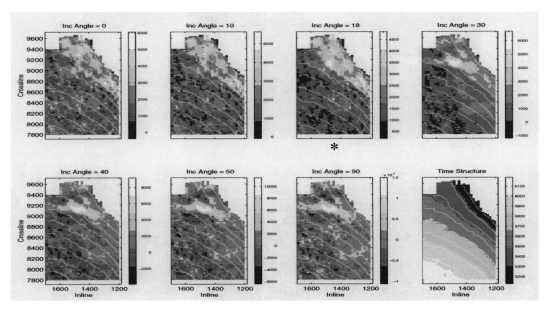

Fig. 5.16   Amplitude maps at different angles over the Auger Field in the Gulf of Mexico. The optimum coincidence of structure and amplitude anomaly is at an angle of 18°, which is therefore the optimum fluid stack. Reproduced with permission from Hendrickson (1999).

shale content in the sand, would give much lower values, in the range −1.5 to −3. The slope of the trend observed in the real seismic trace data is strongly affected by noise. The calculated AVO gradient is very sensitive to random noise on the seismic traces, whereas the intercept is fairly insensitive. Where signal to noise ratio is low (i.e. weak reflectors), the slope on the crossplot will be much higher than the well-based expectation. A complementary approach to this crossplot analysis is to make a series of amplitude maps for different angle ranges and compare the results with the expected behaviour for fluid effects, e.g. amplitude change conforming to structural contours (Hendrickson, 1999; see also fig. 5.16).

A different way of visualising the intercept and gradient data is to create a trace volume of $R_0 * G$ values. This is useful where pay sand has a class III response but brine sand is class I. Pay sand will then have a positive $R_0 * G$ value and brine sand a negative one. By using a colour bar that accentuates the positive values, pay sand is easily identified in the $R_0 * G$ volume (fig. 5.17).

The presence of AVO effects can be a problem for well-to-seismic ties if synthetic seismograms are constructed for the zero-offset case. In principle, well synthetics should

Intercept*Gradient

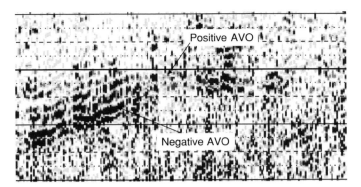

Fig. 5.17 Intercept ∗ gradient section. Positive values are displayed in red and indicate pay sand.

be calculated for a range of incidence angles and stacked to give a trace that can be compared with the real seismic data. In practice the match from seismic to zero-offset synthetic is often reasonable for class I or class III reflectors. It is with class IIp events that the worst problems arise, as the stacked amplitude may be very small and perhaps opposite in sign to the zero-offset response. A poor tie to the zero-offset synthetic is then inevitable.

A complication for AVO analysis is the effect of tuning. Because the TWT difference between two closely spaced reflectors in a gather will vary with incidence angle (before NMO correction), it follows that tuning effects will vary across the gather and distort the AVO response in ways that are hard to recognise except close to a well where bed thicknesses are known and tuning can be modelled.

5.5 Rock physics for seismic modelling

To understand observed amplitude effects, we often need to know how rock densities and seismic velocities (both P and S) are affected by fluid fill (brine, oil or gas), by porosity, by pressure, by clay content, and so on. This is a large subject. A detailed account can be found in Mavko *et al.* (1998), for example, and we shall only provide a summary here. The methods we shall describe work best for medium- to high-porosity sandstones, and are applicable to carbonates only when the pore structure is relatively uniform with a pore size very much smaller than the sonic wavelength. Problems arise in the case of rocks having low porosity and permeability; velocities recorded by the sonic log may be different from those applicable to surface seismic data, because of dependence of seismic velocity on frequency (*dispersion*).

5.5.1 Fluid effects

In general, oil or gas fill will reduce the P velocity significantly compared with the brine case, and for gas the effect is often fully developed at saturations of a few percent (fig. 5.18(a)). With increasing gas saturation beyond this point, the lowering of density becomes important and the seismic velocity starts to increase again. The density decreases linearly as gas saturation increases. The combined effect on acoustic impedance is illustrated in fig. 5.18(b). The impedance of gas sands drops sharply from the brine case for gas saturations of a few percent, and then decreases almost linearly as gas saturation increases. Thus, low gas saturations may give reflections bright enough to be confused with commercially significant accumulations. The effect of oil is more linear over the entire saturation range, with little effect at low oil saturation but an often strong effect at high saturations. S velocity is only slightly affected by differences in fluid fill, via the effect on density; the S velocity is slightly higher for the oil and gas cases.

The effect of fluid fill on P and S velocities can be calculated using Gassmann's (1951) equations. They are applicable at seismic frequencies to rocks with intergranular porosity and fairly uniform grain size, and describe how the bulk and shear modulus of a rock are related to the fluid fill.

The bulk modulus is a measure of resistance to change in volume under applied stress, and the shear modulus is a measure of resistance to change in shape. P and S velocities are related to the bulk modulus K, shear modulus μ and density ρ by the equations:

$$V_{\mathrm{p}} = \sqrt{\frac{K + \frac{4\mu}{3}}{\rho}}$$

and

$$V_{\mathrm{s}} = \sqrt{\frac{\mu}{\rho}}.$$

Gassmann's equations assert that the bulk modulus (K_{sat}) of the rock saturated with a fluid of bulk modulus K_{fl} is given by

$$\frac{K_{\mathrm{sat}}}{K_{\mathrm{ma}} - K_{\mathrm{sat}}} = \frac{K_{\mathrm{d}}}{K_{\mathrm{ma}} - K_{\mathrm{d}}} + \frac{K_{\mathrm{fl}}}{\phi(K_{\mathrm{ma}} - K_{\mathrm{fl}})},$$

where K_{ma} is the bulk modulus of the matrix material, K_{d} is the bulk modulus of the dry rock frame, and ϕ is the porosity. The analogous relation for the shear modulus is given by Gassmann as

$$\mu_{\mathrm{sat}} = \mu_{\mathrm{d}}.$$

This means that the shear modulus is the same irrespective of fluid fill. This is intuitively reasonable, as all fluids have zero shear modulus and are equally unable to help to resist changes in shape of the rock under an applied stress.

(a)

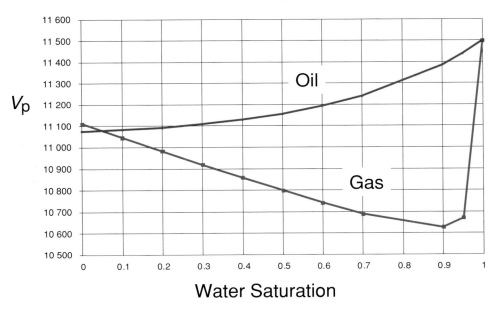

(b)

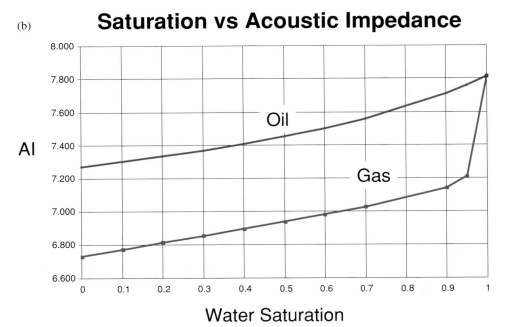

Fig. 5.18 (a) Calculated curves of P velocity (ft/s) versus water saturation for oil and gas cases in an example sandstone; (b) calculated curves of P impedance (km/s·g/cm^3) versus water saturation for oil and gas cases. In real rocks, a water saturation less than 0.15 is unlikely.

At first sight these equations are not very useful, because to calculate the saturated moduli we need to know the moduli for the dry rock frame. It may be possible to measure dry rock properties directly in the laboratory, or predict them from porosity and mineralogy using various theoretical models. However, laboratory measurements are not often available, and the theoretical predictions are quite sensitive to the shape of the pores, which may not be known. This problem can be side-stepped if all we want to do is to calculate the moduli for some particular fluid fill (e.g. gas or oil) when we know the saturated moduli for some other fluid fill (e.g. brine). If we have a well with wireline log data, then the initial saturated moduli can be calculated from the P and S sonic and density logs. Sometimes, however, the log data will be of doubtful quality. A useful check is possible because the Gassmann calculation proceeds via calculation of the parameters for the dry rock frame, and these can be compared with what is generally expected for the particular rock type. The workflow to do this is as follows.

(1) Calculate the shear modulus from the measured shear velocity and density. If there is no shear velocity log, it can be predicted from other log data by methods described below:

$$\mu = V_s^2 \rho.$$

(2) Calculate the saturated bulk modulus from

$$K_{sat} = V_p^2 \rho - \frac{4\mu}{3}.$$

(3) Derive the dry bulk modulus from

$$K_d = \frac{K_{sat}\left(\frac{\phi K_{ma}}{K_{fl}} + 1 - \phi\right) - K_{ma}}{\frac{\phi K_{ma}}{K_{fl}} + \frac{K_{sat}}{K_{ma}} - 1 - \phi}.$$

(4) For QC purposes, calculate the dry rock Poisson ratio from

$$\sigma_d = \frac{3K_d - 2\mu}{2\mu + 6K_d}.$$

The dry bulk modulus and Poisson's ratio can be compared with expected values for particular types of rock, as explained shortly. It may be necessary to edit the data to make sure that the values remain within a reasonable range.

(5) When the values in step (4) are acceptable, calculate the fluid moduli for the new case. Here and in step (3) above we need to be able to estimate the moduli of mixtures of fluids, i.e. hydrocarbons and brine. The moduli for the individual constituents can be obtained by methods to be considered below, and combined using the equation

$$\frac{1}{K_{fl}} = \frac{S_w}{K_w} + \frac{1 - S_w}{K_h},$$

where S_w is the water saturation (decimal fraction), K_h is the bulk modulus of the hydrocarbon and K_w is that of brine. This formula shows why even a small amount of gas (which will have a very low K_h) causes a large decrease in K_{fl}, resulting in a large decrease in V_p.

(6) Calculate ρ_{fl} and ρ_b, the fluid and bulk densities for the new case, using

$$\rho_{fl} = \rho_w S_w + (1 - S_w)\rho_h$$

and

$$\rho_b = \rho_{ma}(1 - \phi) + \phi\rho_{fl}.$$

(7) Determine K_{sat} using

$$K_{sat} = K_d + \frac{\left(1 - \frac{K_d}{K_{ma}}\right)^2}{\frac{\phi}{K_{fl}} + \frac{1-\phi}{K_{ma}} - \frac{K_d}{K_{ma}^2}}.$$

(8) Derive V_s using

$$V_s = \sqrt{\frac{\mu}{\rho_b}}.$$

(9) Derive V_p from

$$V_p = \sqrt{\frac{K_{sat} + 4\mu/3}{\rho_b}}.$$

We now address some of the detailed issues in these calculations.

5.5.1.1 Calculating fluid parameters

Equations to calculate fluid properties have been given by Batzle & Wang (1992). Properties are dependent on pressure and temperature. For brines, the salinity has a significant effect on the bulk modulus (high salt concentration imparts greater stiffness). For oils, the properties depend on API gravity and gas–oil ratio. If data are available from analysis of oil samples, then it may be possible to use directly measured values of bulk modulus and density. (Note, however, that for our calculation we need the adiabatic bulk modulus, and reservoir engineers usually measure an isothermal modulus, so a correction needs to be applied.) For gas, the properties depend on the specific gravity, which reflects the concentration of molecules heavier than methane. Oil-case values are always intermediate between the brine and gas cases, but exactly where they fall depends on the oil concerned: low API and GOR oils will be close to the brine case, whereas high API and GOR oils will be close to the gas case. Figure 5.19 shows some representative curves for North Sea data. The interplay of temperature and pressure effects means that the shapes of these curves are not intuitive, and for accurate work the fluid properties need to be calculated from the Batzle & Wang equations.

Fluid Properties
(North Sea Example)

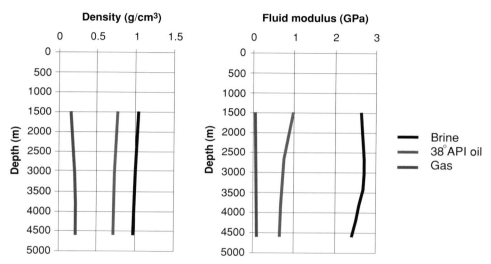

Fig. 5.19 Typical values of density and fluid modulus as a function of depth in the North Sea.

5.5.1.2 Calculating matrix parameters

Elastic moduli and densities for individual minerals can be taken from published values such as those shown in fig. 5.20. Where there is more than one mineral in a rock then a weighted average has to be calculated of the parameters. For density this is a straightforward weighting by fractional composition. For the bulk modulus, there are several possible approaches (Mavko *et al.*, 1998). The simplest is the Voigt–Reuss–Hill average, which calculates the average of the arithmetic and harmonic weighted average values. Thus, for a quartz–clay mixture we would calculate a Voigt bulk modulus:

$$K_v = K_{qtz} V_{qtz} + K_{cl} V_{cl}$$

and a Reuss modulus:

$$K_r = \cfrac{1}{\cfrac{V_{qtz}}{K_{qtz}} + \cfrac{V_{cl}}{K_{cl}}}$$

and then

$$K_{vrh} = \frac{K_v + K_r}{2},$$

where V is the volume fraction of the constituent denoted by the suffix. For the case above, the clay volume fraction would be taken as the shale volume fraction derived from petrophysical analysis.

Examples of Solid Mineral Elastic Parameters

	Density	Bulk Modulus	Shear Modulus	V_p	V_s	Poisson's Ratio	
Quartz	2.65	36.6	45	6.0376	4.1208	0.064	
Chert	2.35	26	32	5.4	3.7	0.058	
Calcite	2.71	76.8	32	6.6395	3.4363	0.317	
Dolomite	2.87	94.9	45	7.3465	3.9597	0.295	
Aragonite	2.94	47	39	5.8	3.6	0.187	
Magnesite	3.01	114	68	8.2	4.75	0.247	
Na-Felspar	2.62	55	28	5.9	3.3	0.272	
K-Felspar	2.56	48	24	5.6	3.05	0.289	
Ca-Felspar	2.73	85	38	7.05	3.75	0.303	
Clays (approx)	2.68	41	17	4.9	2.5	0.324	
Muscovite	2.82	52	31.5	5.8	3.35	0.25	
Biotite	3	50	27.5	5.4	3	0.277	
Halite	2.16	25.2	15.3	4.6	2.65	0.252	
Anhydrite	3	66.5	34	6.15	3.4	0.28	
Gypsum	2.31	58	30	6.75	3.7	0.285	
Pyrite	5.02	158	149	8.4	5.45	0.137	

Density in g/cm³, elastic moduli in GPa, velocities in km/s.

Fig. 5.20   Mineral properties, after Simmons & Wang (1971).

5.5.1.3 Invasion effects

Unfortunately, the density and sonic logs recorded in the borehole may not record values truly representative of the formation. The pressure in the borehole is usually kept higher than the fluid pressure in the formation being drilled. This pressure differential forces drilling fluid into permeable formations, replacing the original fluid in an *invaded zone* around the borehole. Many different types of drilling mud are in use. Water-based muds may contain water of any salinity from very low to nearly saturated. Forcing this water into an oil-bearing reservoir will of course change its density and seismic velocity. Similarly, if an oil-based mud is used, it will change the properties of a brine-filled reservoir in the invaded zone. The extent of the invaded zone depends on several factors, including the permeability of the formation and the length of time that the borehole is left exposed to circulating fluids. The effect on the sonic log also depends on the source–receiver spacing; the wider the spacing, the more likely it is that the tool will measure values in the formation beyond the invaded zone. It is possible to use resistivity logs to estimate the oil saturation in an invaded zone, and then use steps (5) and (6) of section 5.5.1 to estimate the fluid properties in the invaded zone, given the oil and water properties. Gassmann substitution can then be used to correct logs back to what they would be if the formation fluid had not been disturbed. In practice, this may not be a reliable approach, given that the logs may or may not be reading in the invaded zone, depending on the distance that invasion has penetrated away from the borehole. If different fluid zones of the same reservoir have been drilled, perhaps with various mud types in different wells, then a useful check is possible. Starting from a brine-filled reservoir drilled with water-based mud, Gassmann substitution is used to predict the effect of oil or gas saturation, and these values are then compared with the observed logs in oil zones drilled with a water-based mud or brine zones drilled with an oil-based mud to see whether invasion is causing problems. This only works if the reservoir properties (porosity, clay content) are similar for the various zones to be compared; it is unfortunately quite common for hydrocarbon fill to influence diagenesis and so porosity.

5.5.2   P-wave velocity and porosity

Porosity exerts a strong influence on P velocity. A relationship that is often used to describe the effect in brine-filled sands is the Wyllie time-average equation (Wyllie *et al.*, 1958):

$$\frac{1}{V} = \frac{\phi}{V_{\text{fl}}} + \frac{1 - \phi}{V_{\text{ma}}},$$

where V is the velocity in a rock of matrix velocity V_{ma} and porosity ϕ containing fluid of velocity V_{fl}. Despite the similarity to the equation for calculating densities, it is important to note that the Wyllie equation has no proper physical basis and is essentially an empirical result. It is in effect an upper bound for velocity at any given porosity. Velocities are affected by the degree of consolidation and the geometry of the pores. The equation is applicable for consolidated sands in which small cracks and flat pores have been closed, and predicts a strong effect of porosity on velocity (fig. 5.21).

5.5.3   P-wave velocity and clay content

The presence of clay has a significant effect on the properties of a sandstone. The details depend on how the clay is distributed in the rock. There is a conflict between the effects on the matrix (softening it and so reducing velocity) and the effect on the porosity (stiffening it and increasing velocity). Experiments by Marion *et al.* (1992) showed that introducing clay into an experimental sand led to an increase in velocity due to the change in porosity. On the other hand, fig. 5.22 shows that a large velocity reduction is caused by a small amount of clay, at a given porosity. This plot was created as a particular case of the modelling methodology due to Xu & White (1995). Such effects are observed in actual examples.

5.5.4   P-wave velocity and density

In cases where a sonic log is available but density is not (or vice versa), it is useful to be able to predict one from the other. Some general lithology-specific empirical relations

Sandstone Porosity vs Compressional Velocity

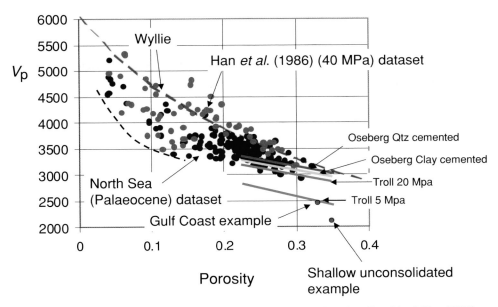

Fig. 5.21 Sandstone compressional velocity vs porosity. Trends based on Dvorkin & Nur (1995).

Effect of Clay on Compressional Velocity

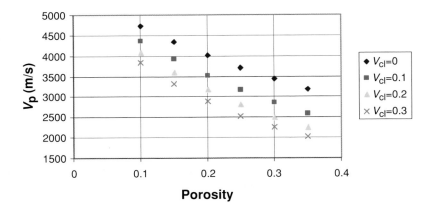

Output from Xu–White (1995) model

Fig. 5.22 Effect of clay on compressional velocity (Xu–White model).

$$\text{Density} = aV_{\text{p}}^{f}$$

Lithology	a	f
ss/sh avg	1.741	0.25
shale	1.75	0.265
sandstone	1.66	0.261
limestone	1.5	0.225
dolomite	1.74	0.252
anhydrite	2.19	0.16

Fig. 5.23   P velocity–density relations. V is in km/s, density in g/cm^3.

have been given by Castagna (1993), as shown in fig. 5.23. An improved version of the limestone relation has been given by Mavko *et al.* (1998):

$$\rho = 1.359V_{\text{p}}^{0.386},$$

where the density is in g/cm^3 and the velocity is in km/s.

Where there exist some wells with both sonic and density data for an interval being studied, it is better to use them to construct a specific relationship rather than use these general equations.

5.5.5   Shear velocity

Ideally, shear velocity should be determined by direct measurement. However, reliable shear velocity logging is a fairly recent introduction, and many wells have P but not S wave logs. It is therefore useful to be able to predict S velocity from P velocity, for use as an input to fluid substitution and AVO modelling. This is also a useful check on measured shear logs, which are sometimes of doubtful quality; shear log processing is to some extent interpretive. Also, some types of sonic tool are not able to record shear velocities lower than the compressional velocity in the drilling mud.

Various empirical relations have been proposed. A useful method is described by Greenberg & Castagna (1992). V_{s} is predicted from V_{p} and mineral fractions using four V_{p}–V_{s} relations:

sand $V_{\text{s}} = 0.804\,16V_{\text{p}} - 0.855\,88$
shale $V_{\text{s}} = 0.769\,69V_{\text{p}} - 0.867\,35$
limestone $V_{\text{s}} = -0.055\,08V_{\text{p}}^{2} + 1.016\,77V_{\text{p}} - 1.030\,49$
dolomite $V_{\text{s}} = 0.583\,21V_{\text{p}} - 0.077\,75$

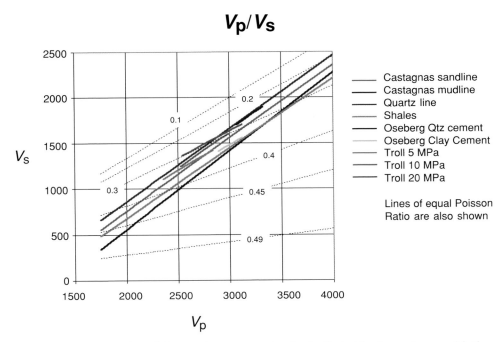

Fig. 5.24 Schematic plots of shear against compressional velocity (m/s) using various empirical relations for a range of lithologies. Lines of constant Poisson's ratio are also shown.

where the velocities are in km/s. The shear velocity for the rock is calculated as the average of the weighted arithmetic and harmonic averages:

$$V_{\text{arith}} = \sum V_{s_i} w_i$$

and

$$V_{\text{harm}} = \frac{1}{\sum \frac{w_i}{V_{s_i}}}$$

where the summation is across the four possible constituents, each with a volume fraction w and shear velocity V_s.

Some typical schematic plots of V_s against V_p are shown in fig. 5.24. The quartz line, $V_s = 0.8029V_p - 0.7509$, was derived by one of the authors (R. W. S.) from the work of Murphy *et al.* (1993) and other data and predicts higher V_s than the Castagna (1993) relations. With increasing velocity (in general, decreasing porosity), Poisson's ratio decreases. For V_p below about 2500 m/s, none of these relations is applicable; they tend to predict too high a shear velocity, perhaps by a wide margin.

5.5.6   Dry rock moduli

As we saw above, dry rock moduli are calculated as a step in performing Gassmann fluid substitution. Comparing them with expected values is a useful check on the accuracy of the input data.

Dry rock moduli vary systematically with porosity. Murphy *et al.* (1993) derived the following empirical relations for clean sands from laboratory measurements:

$$K_d = 38.18(1 - 3.39\phi + 1.95\phi^2)$$
$$\mu = 42.65(1 - 3.48\phi + 2.19\phi^2),$$

where ϕ is the fractional porosity and the moduli are in GPa.

A different approach, though leading to a similar outcome, is the critical porosity model (Nur *et al.*, 1998). In this model, dry rock elastic moduli vary linearly from the mineral value at zero porosity to zero at the critical porosity, where the rock begins to behave as though it were a suspension (fig. 5.25). The value of the critical porosity depends on grain sorting and angularity, but for sands is usually in the range 35–40%. The predictions can be unreliable at high porosities, as the critical value is approached. Figures 5.26 and 5.27 show some examples of plots of elastic moduli against porosity for real rocks. There are variations in the moduli of real sandstones due to the presence of clay and microcracks. Many points plot well below the model predictions, which are only really applicable to consolidated sands. Dry rock Poisson's ratios are generally in the range 0.1–0.25 for sandstones (fig. 5.28).

For carbonates, the picture is more complicated. Chalks are amenable to the same general approach as sandstones, but in most carbonates the moduli are strongly affected by pore shape. Marion & Jizba (1997) give some examples.

The Critical Porosity Model

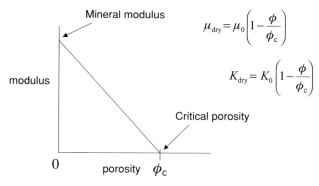

$$\mu_{dry} = \mu_0 \left(1 - \frac{\phi}{\phi_c}\right)$$

$$K_{dry} = K_0 \left(1 - \frac{\phi}{\phi_c}\right)$$

Fig. 5.25   The concept of critical porosity.

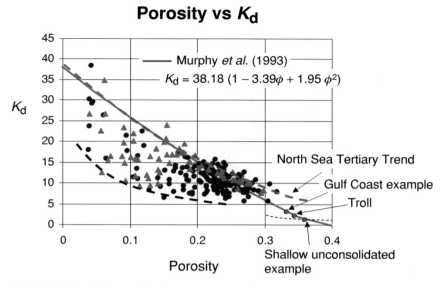

Fig. 5.26 Plot of observed dry bulk modulus against porosity for some real sandstone examples. Broken black line = empirical lower bound for these data.

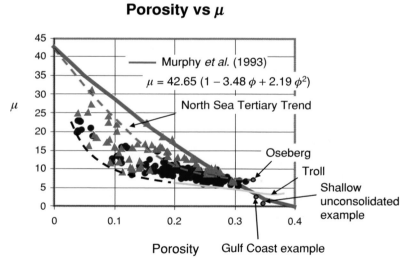

Fig. 5.27 Plot of observed shear modulus against porosity for some real sandstone examples. Broken black line = empirical lower bound for these data.

5.6   Assessing significance

In all but the simplest cases, assessing the significance of an amplitude anomaly (or the lack of one where it might be expected) is difficult. When investigating an

Porosity vs Dry Rock Poisson Ratio

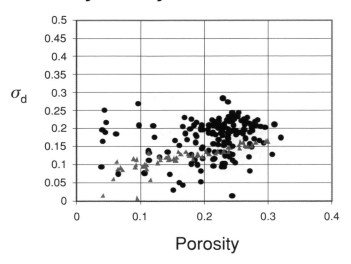

Fig. 5.28   Dry rock Poisson ratio versus porosity for some real sandstone examples.

amplitude change, or a flat event that may be a DHI, the following checklist may be useful.

Are there analogues to show the effects of lithology/ fluid fill on seismic amplitude?
Are there enough good quality well data to carry out modelling studies?
Are the well data from a closely analogous case (e.g. sedimentary environment)?
Is well data quality good enough (e.g. correction of logs for invasion effects)?
Is the amplitude change predicted from the model big enough to be detectable?
Has modelling addressed enough scenarios, including non-pay cases?
Is the effect of change in fluid greater than the effect of likely porosity changes?
Is there a good well tie to establish seismic phase and polarity?
Alternatively, can phase and polarity be deduced from known isolated reflectors?
Are horizon picks uncertain?
Has seismic been processed to preserve amplitudes?
In particular, has short-gate (<1000 ms) AGC been avoided?
Does the seismic have good signal/noise and imaging?
Are multiples a problem?
Do overburden complications affect seismic propagation?
Are the pre-stack data good enough to use AVO methods?
If there is a flat spot, does it have the right polarity?
Is the flat spot discordant with bedding-plane reflectors?
Is there a bright spot/dim spot?
Is the degree of brightening/dimming as expected for a fluid effect?
Are several features concordant in indicating hydrocarbons?

Is there adequate data coverage to map the features?

Are the features consistent from line to line?

Do the features show conformance with depth structure?

Are the effects consistent with analogues?

Fluid effects are sometimes more visible on seismic than Gassmann modelling suggests we have a right to expect; for this reason, we should always look for structure-conforming amplitude change in the real seismic data, even where modelling says that the effects may be too small to see. It is sometimes possible to see the effect of hydrocarbons indirectly, because they tend to preserve porosity during diagenesis. The biggest problem in understanding amplitude anomalies is the need to bear in mind a rich enough set of possible explanations. In particular, it is hard to anticipate effects due to lateral lithological change different from any recorded in existing wells in the area; generation of a range of geological models, based on analogues, is the key to exploring the possibilities.

References

Aki, K. & Richards, P. G. (1980). *Quantitative Seismology*. W.H. Freeman and Co.

Batzle, M. L. & Wang, Z. (1992). Seismic properties of pore fluids. *Geophysics*, **57**, 1396–408.

Campbell, S. J. & Gravdal, N. (1995). The prediction of high porosity chalks in the East Hod field. *Petroleum Geoscience*, **1**, 57–70.

Castagna, J. P. (1993). Rock physics – the link between rock properties and AVO response. In: Offset dependent reflectivity – theory and practice of AVO analysis (eds. J. P. Castagna & M. M. Backus), *Investigations in Geophysics*, **8**. Society of Exploration Geophysicists.

Castagna, J. P. & Swan, H. W. (1997). Principles of AVO crossplotting. *The Leading Edge*, **16**, 337–42.

Dobrin, M. B. (1976). *Introduction to Geophysical Prospecting*. McGraw-Hill.

Dvorkin, J. & Nur, A. (1995). Elasticity of high porosity sandstones: theory for two North Sea datasets. *Geophysics*, **61**, 1363–70.

Gardner, G. H. F., Gardner, L. W. & Gregory, A. R. (1974). Formation velocity and density: the diagnostic basics for stratigraphic traps. *Geophysics*, **39**, 770–80.

Gassmann, F. (1951). Elastic waves through a packing of spheres. *Geophysics*, **16**, 673–85.

Greenberg, M. L. & Castagna, J. P. (1992). Shear wave velocity estimation in porous rocks: theoretical formulation, preliminary verification and applications. *Geophysical Prospecting*, **40**, 195–209.

Gregory, A. R. (1977). Aspects of rock physics from laboratory and log data are important to seismic interpretation. In: Seismic stratigraphy – applications to hydrocarbon exploration. *AAPG Memoir* **26** (ed. C. E. Payton).

Han, D.-H., Nur, A. & Morgan, D. (1986). Effects of porosity and clay content on wave velocity in sandstones. *Geophysics*, **51**, 2093–107.

Hendrickson, J. (1999). Stacked. *Geophysical Prospecting*, **47**, 663–706.

Marion, D., Nur, A., Yin, H. & Han, D. (1992). Compressional velocity and porosity of sand–clay mixtures. *Geophysics*, **57**, 554–63.

Marion, D. & Jizba, D. (1997). Acoustic properties of carbonate rocks. In: Carbonate seismology (eds. I. Palaz & K. J. Markurt), *Geophys. Dev. Series*, **6**. Society of Exploration Geophysicists.

Mavko, G., Mukerji, T. & Dvorkin, J. (1998). *The Handbook of Rock Physics*. Cambridge University Press.

Murphy, W., Reischer, A. & Hsu, K. (1993). Modulus decomposition of compressional and shear velocities in sand bodies. *Geophysics*, **58**, 227–39.

Nur, A., Mavko, G., Dvorkin, J. & Galmudi, D. (1998). Critical porosity: a key to relating physical properties to porosity in rocks. *The Leading Edge*, **17**, 357–62.

Ross, C. P. & Kinman, D. L. (1995). Nonbright-spot AVO: two examples. *Geophysics*, **60**, 1398–1408.

Rutherford, S. R. & Williams, R. H. (1989). Amplitude versus offset variations in gas sands. *Geophysics*, **54**, 680–8.

Shuey, R. T. (1985). A simplification of the Zoeppritz equations. *Geophysics*, **50**, 609–14.

Simmons, G. & Wang, H. (1971). *Single Crystal Elastic Constants and Calculated Aggregate Properties*. Michigan Inst. of Tech. Press, Cambridge, Mass.

Wyllie, M. R. J., Gregory, A. R. & Gardner, G. H. F. (1958). An experimental investigation of factors affecting elastic wave velocities in porous media. *Geophysics*, **28**, 459–93.

Xu, S. & White, R. E. (1995). A new velocity model for clay–sand mixtures. *Geophysical Prospecting*, **43**, 91–118.

Zoeppritz, K. (1919). Uber reflexion und durchgang seismischer Wellen durch Unstetigkerlsflaschen. *Uber Erdbebenwellen VIIB. Nachrichten der Koniglichen Gesellschaft der Wissenschaften zu Gottingen, Math. Phys.*, **K1**, 57–84.

Inversion

The fundamental idea of inversion to seismic impedance is very simple. A reflectivity seismic section contains reflections that can be studied by the methods discussed in chapters 3–5. These reflections show where there are changes in acoustic impedance in the subsurface. Inversion is the process of constructing from this reflectivity dataset a section that displays the acoustic impedance variation in the subsurface directly. As we shall see, this often makes it easier to interpret the data in geological terms, because it focuses attention on layers and lateral variations within them, rather than on the properties of the interfaces between layers that cause the seismic reflections. This is an idea that has been known for many years (see, for example, Lindseth, 1979) but has not been used very much until recently, probably because good results require input reflectivity data of excellent quality; the availability of modern 3-D datasets has triggered an upsurge of interest in the technique.

This chapter begins with a summary of the principles, then discusses some practical processing workflows with particular attention to the issues that are critical for the quality of the results, continues with some practical examples to demonstrate the benefits of the technique, and concludes with a summary of some specialised advanced applications.

6.1 Principles

A simple model for zero-offset seismic response is illustrated in fig. 6.1. The subsurface is represented as a number of layers, each with its own acoustic impedance A; the zero-offset reflection coefficient at the interface between the nth and the $(n + 1)$th is given by

$$R_n = (A_{n+1} - A_n)/(A_{n+1} + A_n)$$

or for a small change in impedance δA, then

$$R = \delta A/2A = 0.5\delta(\ln A).$$

The reflection coefficient series is convolved with the seismic wavelet to give the seismic trace. Inversion aims to start from the seismic trace, remove the effect of the

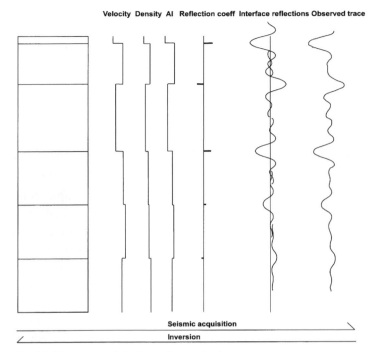

Velocity Density AI Reflection coeff Interface reflections Observed trace

Seismic acquisition

Inversion

Fig. 6.1   Principle of the inversion method.

wavelet to get back to the reflection coefficient series, and from them derive the layer impedances. It has to assume that the starting seismic data are free from correlated noise (e.g. multiples). Also, the wavelet present in the data has to be estimated; many inversion methods derive the wavelet from a well tie and have to assume that it does not change laterally away from the well. There is also an amplitude calibration to be taken into account; real seismic traces are not directly output as reflection coefficient values, but are scaled to give some convenient but arbitrary rms average over a trace. At least over a limited time-gate, the ratio of reflection coefficient to trace amplitude has to be constant if inversion is going to work. Care is needed during seismic processing to avoid steps that might introduce artificial amplitude changes vertically or horizontally. However, locally variable effects in the overburden (e.g. shallow gas) can reduce the seismic energy penetrating to deeper reflectors and so reduce the reflected signal; left to itself, the inversion process would try to interpret this as a decrease in impedance contrast across the deeper interfaces. To remove such artefacts, a long-gate AGC may be applied, which scales amplitudes so as to remove lateral variation when averaged over a TWT interval of 1 s, for example.

Another issue is that the seismic traces contain data of limited bandwidth; the frequencies present depend on the rock properties and the seismic acquisition technique, but might be in the range from 5 to 50 Hz. This means that the low frequencies in particular, which are critical for the estimation of absolute impedance values, cannot be obtained from the seismic data, but have to be added from elsewhere, usually from a model based on well data and geological knowledge.

6.2   Procedures

6.2.1   SAIL logs

A very simple approach to inversion was described by Waters (1978). This is the Seismic Approximate Impedance Log (SAIL), which is derived as follows. Instead of the discrete reflection coefficients of section 6.1, we define a piecewise continuous function of reflection time t which is called the reflectivity:

$$r(t) = \lim_{\delta t \to 0} \delta R / \delta t = 0.5 \mathrm{d}(\ln A)/\mathrm{d}t$$

from which we see that

$$\ln \frac{A(t)}{A(0)} = 2 \int_{t_0}^{t} r(t)\mathrm{d}t$$

so that

$$A(t) = A(0) \exp \left(2 \int_{t_0}^{t} r(t)\mathrm{d}t \right).$$

What we actually have is a seismic signal $s(t)$. If we can assume that the wavelet is zero-phase and the data are noise-free, then the signal s is a band-limited representation of $r(t)$, related to it by an unknown scaling factor, so we can write

$$A(t) = A(0) \exp \left(\alpha \int_{t_0}^{t} s(t)\mathrm{d}t \right),$$

where α takes account of this unknown scaling. Assuming that the exponent is small, as is likely in practice, the exponential term can be expanded as a series and truncated, to give the approximate relation:

$$A(t) = A(0) \left\{ 1 + \alpha \int_{t_0}^{t} s(t)\mathrm{d}t \right\}$$

or

$$(A(t) - A(0))/A(0) = \alpha \int_{t_0}^{t} s(t)\mathrm{d}t,$$

which allows us to calculate fractional change in impedance as a function of time by integrating reflectivity from a starting time t_0 at which we know the impedance $A(0)$; the unknown scaling constant prevents us from turning this into absolute values of impedance. The SAIL trace results will clearly be limited to the seismic bandwidth. Isolated interfaces where there is a large impedance change, such as the top of a massive carbonate within a dominantly shale sequence, will show up as the band-limited version of a step function, which is a bipolar loop. A limitation is the assumption that the seismic wavelet is zero-phase; if at all possible, this needs to be checked by means of a well tie.

A particularly useful formulation of this approach can be derived by thinking of the reflectivity trace in the Fourier domain, i.e. as the summation of a number of sine waves of different frequencies and phases. Integration of an individual component sine wave of frequency ω gives a sine wave whose phase has been shifted by $90°$ and whose amplitude has been multiplied by $1/\omega$. Integration of the entire trace requires that this process should be applied to every individual sine wave component. Therefore, the seismic trace can be integrated by applying a phase shift of $90°$ and a high-cut filter that falls off as $1/\omega$ across the entire seismic bandwidth, i.e. at 6 dB per octave. Many interpretation software suites contain the functionality to process seismic data in this way, so SAIL traces can be easily generated without the need for any additional processing software.

Figure 6.2 is an example of what can be achieved by this means. It is a seismic line across an area of the Central North Sea, offshore UK. The horizon picked in yellow

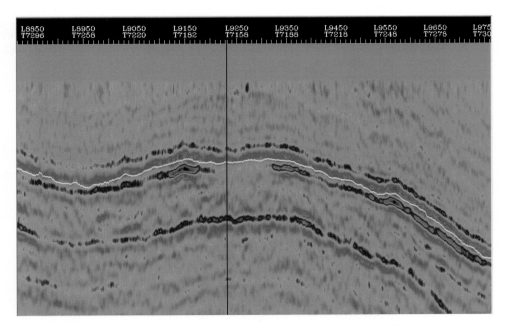

Fig. 6.2 Band-limited inversion result, produced at the workstation using the SAIL methodology.

marks the top of a particular stratigraphic interval. The SAIL section is displayed with a colour bar such that hard intervals are red and soft intervals are violet-purple; the same convention will be used for other impedance sections in this chapter. Patches of soft colours below the yellow marker show the location of soft oil-filled sands. These are actually channels crossing the plane of section, as would be clear from a map view; note the small bumps in the yellow horizon above the soft patches, which are formed because the sandy channel-fill compacts less than the shales on either side as it is buried (*differential compaction*, which often results in sandy intervals being marked by mounded topography). Such a display can be used in a qualitative way, to map out the extent of a hydrocarbon accumulation. It would be less straightforward to try to use the results to make a quantitative prediction of, for example, reservoir porosity, because the traces show only relative impedance. Making a plot of well impedance data, with bandwidth limited to the seismic spectrum, is a useful aid to understanding how the SAIL results relate to the real earth. They often turn out to be surprisingly useful for such a simple technique, because in many cases reservoirs are not thick enough for the missing low-frequency component to be critical.

A refinement of this approach is to try to match the spectrum of the SAIL data to the spectrum of impedance in the real earth. If we compute reflectivity from well log data, we usually find that the earth reflectivity spectrum is not flat over the seismic bandwidth, as is commonly assumed in seismic processing; instead, it has more energy at higher frequencies. By analogy with the optical light spectrum, we can say that the earth reflectivity is not white (i.e. flat) but blue (i.e. higher amplitude at higher frequencies). A first approximation is often that the well log reflectivity has a slope of 3 dB per octave across the seismic spectrum. This can be allowed for by high-cut filtering the 90° phase-rotated trace at 3 dB per octave instead of 6 dB. A more exact approach is to determine the actual reflectivity spectrum from a well log (Lancaster & Whitcombe, 2000).

6.2.2   Extending the bandwidth

To get more information into a seismic section than is actually present in the seismic traces, extra data have to be obtained from elsewhere; different algorithms differ in detail, but they all have the following general features.

(1) Begin with a model that describes the subsurface; explicitly or implicitly, this will contain a number of layers of different acoustic impedance.

(2) Calculate the seismic response from this model, using the wavelet present in the seismic dataset.

(3) Compare the calculated seismic response with the real data.

(4) Modify the model so as to reduce the misfit between the calculated and real seismic, perhaps iteratively and perhaps incorporating constraints on the impedance values that may be assigned to particular layers, on the complexity of the model, and on

the variation of layer parameters from one seismic trace to the next along a seismic line.

(5) Perhaps, add low-frequency information obtained from a model based on geological data.

The additional information that has been taken into account is thus:

(a) the wavelet: removal of its effects is equivalent to a deconvolution, extending bandwidth at the high-frequency end;

(b) the model: the geological input extends bandwidth at the low-frequency end.

To make all this more concrete, it is helpful to work through an actual processing flow. A common approach to building a subsurface model is to split it up into macrolayers, probably several hundreds of ms thick, bounded by the main semi-regional seismic markers, and consisting of a single broad lithology. This makes it easier to construct a geological model to constrain impedance variation within a macrolayer. Inside the macrolayer, the subsurface is represented by means of a series of reflectivity spikes. These spikes, when convolved with the wavelet, should reproduce the observed seismic trace, and integration of the reflectivity will give the impedance variation within the macrolayer. To prevent the software from trying to reproduce all the noise present in the seismic section, it is usual to impose a requirement that the reflectivity spike series is as simple as possible, either in the sense of using a small total number of spikes or using spikes of small total absolute amplitude; the algorithm will trade off the misfit between real and calculated seismic section against the complexity of the spike model, usually under user control of the acceptable degree of misfit. The result is usually called a 'sparse spike' representation of the subsurface.

It is obviously critical to the success of this process that we know the wavelet accurately. This is usually obtained from a well-tie study. As discussed in chapter 3, well synthetic or VSP information will tell us how zero-phase seismic ought to look across the well; comparison with the real data tells us what wavelet is present. Figure 6.3 shows an example display from such a study. To the left, in red, is the candidate wavelet, which in this case is close to being symmetrical zero-phase; to the right is a panel of six (identical) traces showing the result of convolving this wavelet with the well reflectivity sequence derived from log data. They should be compared with the panel of traces in the middle of the figure, which are the real seismic traces around the well location. Various geologically significant markers are also shown. There is clearly a very good match between the real and synthetic data so far as the principal reflections are concerned, so it is possible to have confidence that the wavelet shown is indeed that present in the data. There are also some differences in detail, which will limit the accuracy of an inversion result. These may arise from imperfections in the seismic processing, perhaps the presence of residual multiples or minor imaging problems. With luck, the inversion process will leave some of this low-energy noise out of the inverted image, because of the sparseness of the spike reflectivity series. Another possible source of mismatch is AVO, which will be discussed later in the chapter.

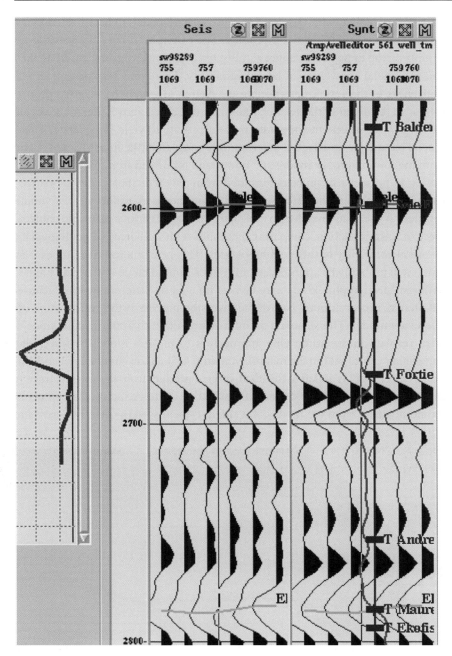

Fig. 6.3 Wavelet, real seismic traces and synthetic traces derived using wavelet.

In many implementations there are further parameters to be adjusted to arrive at an optimal sparse spike inversion product. For example, it may be possible to set constraints on the impedance values within each of the macrolayers, based on the well data and geological knowledge of lateral variability. There may be a parameter that sets the sparseness of the series, determining the degree to which the inverted product tries to follow all the detail of the seismic traces. As discussed above, there is a wavelet scaling parameter that affects the integration of reflectivity to yield impedance. This is often a crucial parameter; in effect, it determines how much of the final inverted section comes from the low-frequency model and how much is derived from the seismic traces themselves. In principle it can be determined from a well tie, but sometimes there are quite large variations in scaling from one well to another, perhaps because of variation in absorption. A useful approach is to look at the results of various plausible choices on a trial piece of seismic section, comparing the results with well data and geological knowledge of lateral variability of different lithologies. This is an interpretive process, and the solution will probably be appropriate only over a limited range horizontally and vertically. Different stratigraphic levels may be best imaged on different inverted datasets.

Often, the most difficult part of the inversion process is the construction of the low-frequency model. This depends on information other than the seismic trace data, very often on well information. One approach is to make a model by interpolating well impedance values within the macrolayers (fig. 6.4). A low-frequency band-limited version of this model is added to the band-limited product from the seismic trace inversion, using the spectrum of the extracted wavelet to decide exactly what frequency range should be taken from the model; this might typically be 0–6 Hz. Where there is good

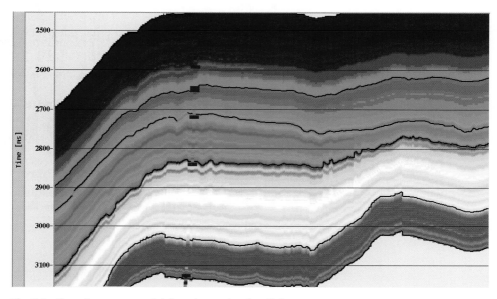

Fig. 6.4 Low-frequency model from interpolated well data.

well control, this process is straightforward; where there are few wells, there is obviously a danger of imposing an incorrect low-frequency trend on the final result. This is particularly true if the wells are drilled in updip locations and the target for the inversion is downdip, with a greater depth of burial causing additional compaction and perhaps containing additional stratigraphic intervals not sampled at all by the well control. It may be possible to allow for compaction effects by putting in trends of increasing impedance with depth, but thick intervals with no well control are bound to cause problems. A last resort if there is no well control is to create the low-frequency model from seismic velocities (stacking or migration). The snag with this, apart from the general issues of the unreliability of seismic velocities as a measure of real rock properties discussed in chapter 3, is that density has to be inferred from velocity in order to calculate impedance; to do this, it is necessary to know the lithology, which may be just the point at issue in an undrilled section.

A section resulting from this process is shown in fig. 6.5. It is created from the same reflectivity data as were used to create fig. 6.2. The effect of the low-frequency information is apparent in the progression from low impedance (purple) at the top of the section to high impedance (yellow/white) at the base. The high-impedance precursor (green/red) to the target horizon (i.e. above the yellow marker) visible in fig. 6.2 is not a feature of fig. 6.5; probably it was an artefact produced by minor departure of the wavelet from the zero-phase assumed in the SAIL processing. However, the definition of the pay sand is not much different from that in fig. 6.2; in this particular case the simple SAIL approach is adequate, partly because the reflectivity data are close to zero-phase and partly because the target layer is the right thickness to have a response within the seismic bandwidth.

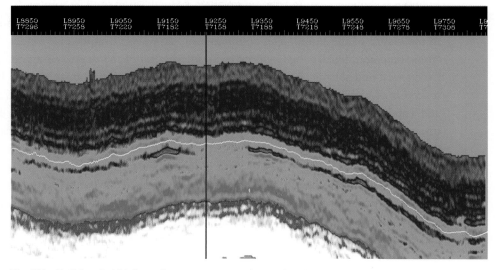

Fig. 6.5   Full-bandwidth inversion over same section as fig. 6.2.

6.3   Benefits of inversion

Several benefits of inversion have been identified, for example by Buiting & Bacon (1999). These include the following.

(1) Impedance displays are layer-oriented, unlike reflectivity displays which are interface-oriented. This greatly facilitates interaction of seismic interpreters with geologists, petrophysicists and reservoir engineers.

(2) Impedance can often be related to reservoir properties such as porosity or net to gross ratio. The inverted dataset can therefore be used directly to constrain reservoir models, to improve volumetric estimates and to target wells more precisely.

(3) Inversion lends itself to stochastic approaches that permit exploration of the range of reservoir models that are consistent with the observed seismic data.

Some examples will help to explain what can be achieved. Although they will be illustrated by means of 2-D sections taken from the 3-D dataset, it is important to bear in mind that the value is much enhanced by the possibility of generating a full 3-D volume of reservoir attributes. This is a requirement for reservoir modelling, but is also a good QC tool. Inversion is interpretive processing, requiring decisions on parameter settings which may rest on geological intuition. A 3-D volume that can be interpreted in a geologically consistent way is good evidence of correct choices; a few 2-D lines may not offer much scope to check whether the parameters give sensible results beyond the traces they were designed on.

6.3.1   Inferring reservoir quality

Figure 6.6 shows a section from an inverted 3-D dataset in the UK Central North Sea. A well was drilled at the location marked by the vertical differently coloured strip in the centre of the figure. The coloured strip shows well impedance values derived from sonic and density logs, and gives an impression of the degree of agreement between well and seismic data. The target for the well was the Fulmar sand, a Jurassic reservoir of variable quality. The other picked events are the Base Cretaceous (Xunc) and Top Zechstein. There is clearly a patch of low-impedance (green) Upper Fulmar reservoir around the well, but most of the Fulmar is harder (red/yellow). These differences can be turned into an estimate of the porosity of the Fulmar sand, using the calibration shown in fig. 6.7. This is a plot of acoustic impedance versus porosity, based on data from several wells; trends can be derived for the prediction of porosity and its uncertainty, as shown for the case of the aggradational facies that forms the low-impedance reservoir in fig. 6.6. In this way, the Fulmar reservoir porosity can be determined from the inversion dataset throughout the 3-D volume. This will improve the estimates of oil in place after a discovery; it will also help the reservoir engineer to assess the quality of reservoir interconnection, which is important for

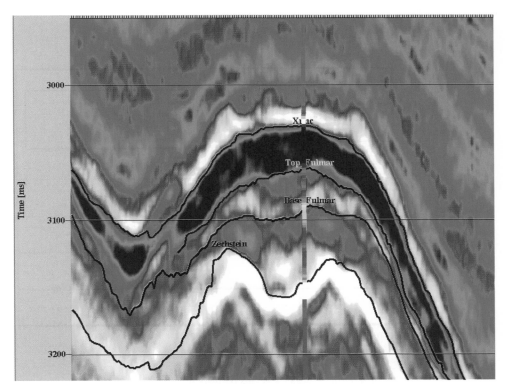

Fig. 6.6   Section from an inverted 3-D dataset.

planning development, for example in choosing locations for producers and water-injection wells.

Another example is shown in fig. 6.8. In this case the reservoir is the Upper Jurassic Magnus sand of the UK Northern North Sea. The figure shows the location of a calibration well and the picked top and base of the reservoir sequence. In this case the reservoir unit contains interbedded sands and shales. The upper part of the figure shows an inverted section from the 3-D volume. The sands are hard relative to the shales, and the application of an impedance cutoff allows separation of sands from shales; in the lower part of the figure this has been applied to separate sands (shown in grey) from shales (in pink). The next step would be to estimate porosity in the sands; in this particular study a stochastic approach was used, and the results will be shown after explaining the underlying principles in the next section.

Inversion is also a useful tool for the study of carbonate reservoirs. For example, Story *et al.* (2000) were able to use an inverted dataset to map zones of high and low porosity within a producing reef. Because of its shallow depth (3800 ft), high seismic frequencies could be recorded, and a resolution of 14 ft was achieved in the inverted dataset. Mapping of low-porosity, low-permeability tight zones was critical to an understanding of reservoir behaviour during production; the tight zones turned out

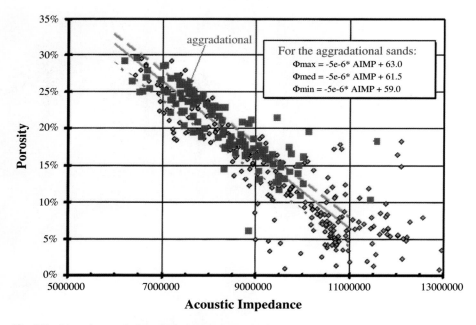

Fig. 6.7   Plot of acoustic impedance versus porosity in Fulmar sands (UK North Sea), based on several wells.

to be discontinuous, making them less effective as baffles to prevent early water influx into producing wells than had been hoped.

6.3.2   Stochastic inversion

Seismic information can be incorporated into a reservoir model by using geostatistical methods; a useful introduction to this topic is provided by Chambers *et al.* (2000). The problem in building a reservoir model is that there is usually only a small number of well penetrations where reliable data on reservoir properties can be obtained directly. To estimate the properties between the wells, it is necessary to understand their spatial variability. As we move away from a well, the values measured there become less and less useful as a way of predicting properties, but we need to know how quickly this fall-off in predictive power happens. Data can be analysed using variograms, which are plots showing how the dissimilarity of a property increases with distance from an observation point; they can be created by comparing properties measured in pairs of wells at different distances apart. The range at which the dissimilarity reaches a constant high value tells us how far away from a well it is reasonable to extrapolate the properties measured there. If this range is, say, several kilometres, then interpolation

Well

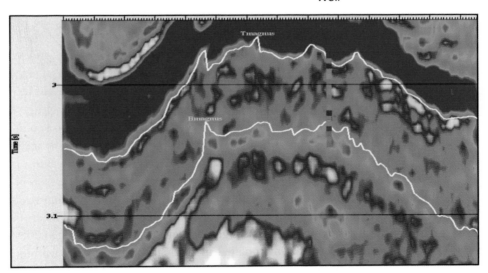

Impedance

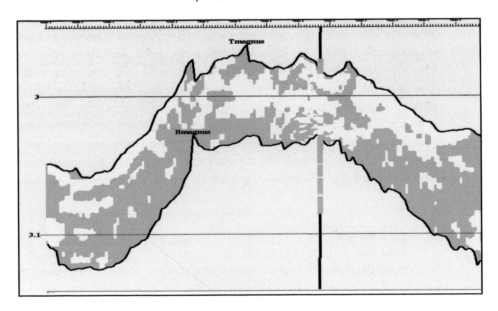

Sand presence

Fig. 6.8 Inverted section and inferred sand/shale separation in Magnus sands, UK North Sea.

between wells may be easy and reliable, whereas if it is only a few hundred metres then there will be problems. Often, the range is different in different directions, being for example greater along strike of a depositional system. Knowing the variogram, a mathematical methodology called kriging allows us to interpolate between the sparse well data in a way that honours the spatial variability of the property. In addition, seismic information can be incorporated by using a methodology called cokriging. For example, impedance from an inverted dataset might be used as an aid to predicting reservoir porosity. Cokriging would produce a map that would honour the 'hard' data at the wells exactly, and would infer values between the wells based on a combination of kriged interpolation between them and the 'soft' evidence provided by the impedance data. However, the resulting map is a smooth map of most likely values; it does not capture the true variability within the reservoir, because as we have seen in chapter 4 the spatial resolution of a seismic dataset is limited; inversion will improve resolution vertically but not horizontally. Unfortunately for the reservoir engineer, the flow of fluids in a real reservoir may be strongly influenced by the inhomogeneities that have been smoothed out of the interpolated model. Stochastic inversion offers a way to construct a suite of models ('realisations') of the subsurface that are compatible with the well and seismic evidence and retain the high spatial frequencies.

A way of doing this is outlined by Haas & Dubrule (1994). Suppose that we have a reservoir volume for which we have a reflectivity seismic dataset, and a number of well penetrations (fig. 6.9). At each of four wells in the example, we have an acoustic impedance trace derived from log data. We start by drawing at random a seismic trace location, where we wish to estimate the impedance trace (shown in green in fig. 6.9). A candidate impedance trace is simulated from the well data; in effect, impedance values are drawn at random from a population in such a way that the statistics of the spatial variability of acoustic impedance derived from the well information are honoured. This means that the candidate trace will contain the short-wavelength vertical variation that is found in the well logs. From this impedance trace the reflection response is calculated,

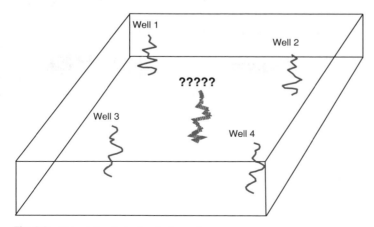

Fig. 6.9 Principle of stochastic inversion.

using the wavelet present in the seismic data. The result is compared with the real seismic trace at the chosen location. If the fit is not satisfactory, the simulation of the impedance trace is repeated using a different random draw from the population, and the comparison with the real data repeated. The simulation is repeated until eventually an impedance trace is found that gives a satisfactory fit to the observed seismic. This is then treated as the true impedance at this location, and is merged with the well data. The entire estimation process is then repeated at another randomly drawn trace location, using the augmented 'well' data from the previous step to estimate impedance at the new location. Eventually, all the trace locations will have estimated impedance traces. The resulting dataset is consistent with the reflectivity data and also with the detailed well information. The solution is not, of course, unique; different final impedance volumes will be found if the random order in which traces are selected for calculation is changed, and if different random draws are made from the impedance population when estimating the trace at any particular point. This means that it is possible to produce a number of different reservoir models, all of which honour the seismic and well information; study of the range of possibilities will quickly show which features of the impedance volume are well constrained (and therefore found in all models) and which are poorly determined (and show high variability from one model to another). Each impedance volume can be used to estimate 3-D porosity or lithology by the methods considered in the previous section.

One simulation produced by applying this technique to the example of fig. 6.8 is shown in fig. 6.10. The scale on the left shows porosity within the sand; shale units

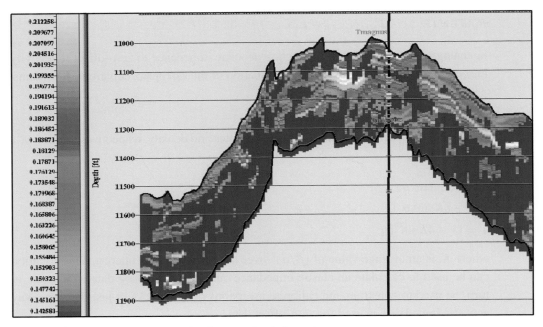

Fig. 6.10   Porosity simulation from stochastic inversion.

within the reservoir are coloured dark blue. The frequency content is obviously much higher than the standard inversion shown in fig. 6.8. As noted above, this result is consistent with the well and seismic data, but there is no guarantee that it is correct; we would need to examine a whole suite of realisations to see which of the features are well constrained. Because of this, the technique tends to be rather time-consuming and expensive, and is generally regarded as a rather specialised tool.

6.4   AVO effects

So far it has been implicitly assumed that the seismic data can be treated as though they were acquired at zero-offset. The discussion of AVO in chapter 5 shows that in some cases there are significant changes of amplitude with offset. In particular, with class III AVO responses it may well give better results to invert a far-offset sub-stack. There is then a problem in that we need somehow to modify the acoustic impedance log at the wells in order to have values that can be compared with far-stack inverted amplitudes. It is not clear what we mean by acoustic impedance at non-zero-offset. It is possible to calculate the reflectivity at each interface and to integrate it to give an 'impedance' variation, but this is not a property of the rock in the same way that true acoustic impedance is; it is a function of the incidence angle.

A possible approach has been suggested by Connolly (1999). He introduces the idea of an elastic impedance which is a function of incidence angle $E(\theta)$, such that the reflection coefficient at an interface is given by

$$R_n(\theta) = (E_{n+1} - E_n)/(E_{n+1} + E_n)$$

in analogy with the equation connecting acoustic impedance to zero-offset reflectivity in section 6.1. From this, using a simplification of the Zoeppritz equations, he obtains

$$E(\theta) = \alpha^x \beta^y \rho^z$$

where α, β and ρ are the P- and S-wave velocities and density, respectively, and x, y, z are given by

$$x = (1 + \tan^2 \theta)$$
$$y = -8K\sin^2 \theta$$
$$z = (1 - 4K\sin^2 \theta),$$

where K is an average value of $(\beta/\alpha)^2$ over the entire zone of interest. This equation can be used to calculate an elastic impedance curve from well log data which can, in turn, be used to constrain and calibrate inversions of sub-stacks in just the same way acoustic impedance is used for zero-offset data.

A benefit of inverting angle stacks to elastic impedance is that a wavelet will be estimated from well ties for each angle range independently. This takes care of the shift to lower frequencies normally seen on the far traces, but more importantly allows the amplitude of the near and far traces to be correctly matched. This removes one of the chief uncertainties in quantitative interpretation of AVO effects. As compared with AVO interpretation on reflectivity data, additional benefits are the reduction of the effect of tuning and the ability to distinguish lateral change in a reservoir from changes in the overburden, just as in acoustic impedance inversion.

Various approaches have been taken to combining the elastic impedance information from different angle ranges to yield rock parameters. Possibilities include the estimation of Poisson's ratio or of compressional and shear elastic moduli, which can be diagnostic of different lithologies or fluid fill. However, the results are sensitive to noise in the data, and careful comparison of the predicted rock parameters with well data is an essential QC step.

References

Buiting, J. J. M. & Bacon, M. (1999). Seismic inversion as a vehicle for integration of geophysical, geological and petrophysical information for reservoir characterisation: some North Sea examples. In: *Petroleum Geology of Northwest Europe: Proceedings of the 5th Conference* (eds. A. J. Fleet & S. A. R. Boldy). Geological Society London.

Chambers, R. L., Yarus, J. M. & Hird, K. B. (2000). Petroleum geostatistics for nongeostatisticians. *The Leading Edge*, **19**, 474–9.

Connolly, P. (1999). Elastic impedance. *The Leading Edge*, **18**, 438–52.

Haas, A. & Dubrule, O. (1994). Geostatistical inversion – a sequential method of stochastic reservoir modelling constrained by seismic data. *First Break*, **12**, 561–9.

Lancaster, S. & Whitcombe, D. (2000). Fast-track "coloured" inversion. *Expanded abstract, SEG Annual Meeting, Calgary.*

Lindseth, R. O. (1979). Synthetic sonic logs – a process for stratigraphic interpretation. *Geophysics*, **44**, 3–26.

Story, C., Peng, P., Heubeck, C., Sullivan, C. & Lin, J. D. (2000). Liuhua 11-1 Field, South China Sea: a shallow carbonate reservoir developed using ultra-high resolution 3-D seismic, inversion, and attribute-based reservoir modelling. *The Leading Edge*, **19**, 834–44.

Waters, K. H. (1978). *Reflection Seismology*. John Wiley, New York.

7   3-D seismic data visualisation

Recently there has been a change in the way that interpreters view seismic data. The traditional method of working, as explained in chapter 3, has been to make different 2-D sections through the 3-D data volume, as inlines, crosslines, random tracks or time slices. The only way to view more than one section at a time was to open multiple windows and view each one in a separate display. Today, largely thanks to relatively low-cost computer power and memory, it is possible to view entire datasets so that the interpreter can quickly get a feel for the actual 3-D nature of the trap. Indeed, several different data volumes can be viewed simultaneously to interrogate various attribute volumes at the same time (fig. 7.1). This has many applications. As shown in fig. 7.1, reflectivity and coherence volumes can be viewed simultaneously when interpreting a fault; this is a way to combine the lateral continuity information from coherency with the identification of the nature of a feature in the standard reflectivity section. Such technology can also be used to view different AVO volumes in the same display, or to examine reflectivity and acoustic impedance (inversion output) volumes at the same time. Different time-lapse seismic volumes (see chapter 8) can be displayed so that production-related changes can be more easily seen.

Also, various types of data can be viewed together with the seismic traces. In fig. 7.2 we see the top reservoir map viewed together with well trajectories and overlain with satellite imagery data of the surface geology. Figure 7.3 is a more traditional combination: the top reservoir horizon together with the 3-D seismic volume from which it was picked. The ability rapidly to scan through the entire seismic data volume together with the reservoir pick is a very quick way of understanding where the picks may require adjusting. For this type of work, display speed is critical. To check the consistency of a pick across a volume of seismic data, the interpreter needs to keep in his head a picture of how the pick looked on the previous view while examining a new one. If it takes several seconds to redraw the display to show a new line, it is hard to maintain this mental picture; the ideal is to have the display change to follow cursor movement on the screen fast enough that the delay caused by the redrawing is not perceptible.

One of the advantages of having the entire data volume loaded in memory is that it is quick to sculpt out different portions of the seismic data. There are many ways to do this. A simple technique is to use interpreted horizons to guide sculpting; for example, data

Fig. 7.1 Different datasets can be viewed simultaneously. Here a volume of standard reflectivity seismic is viewed together with a coherency volume (in blue). Also shown is an interpreted fault plane picked with the aid of both volumes.

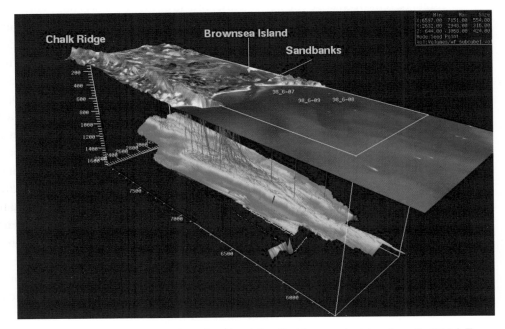

Fig. 7.2 Surface and subsurface data viewed together. In this example, taken from the Wytch Farm oilfield in southern England, we can see the colour-coded reservoir map displayed together with satellite imaging of the surface.

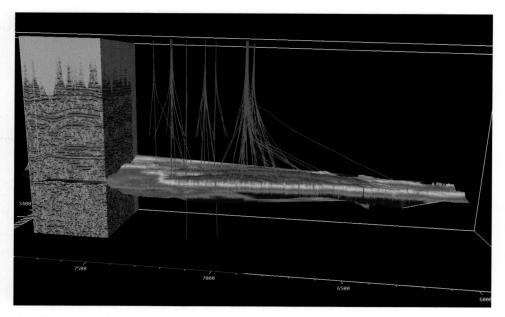

Fig. 7.3 Combination of data types. In this example, we see seismic data together with the top reservoir time map and well trajectories.

might be removed above and below a reservoir so as to leave a 3-D body corresponding to the reservoir alone. A more automated method is sometimes referred to as *voxel picking*. (A voxel is the 3-D equivalent of a pixel, and is the smallest independent unit out of which the 3-D volume is constructed, typically corresponding to a 4 ms time sample and a 12.5 m × 12.5 m horizontal trace grid.) An initial seed point is chosen, and the software then automatically seeks out all connected voxels with amplitudes less (or greater) than a given threshold (fig. 7.4). This is most often applied to acoustic impedance or standard reflectivity volumes, though any seismic attribute that can be calculated at all points in a 3-D volume could be used. For example, we might choose the initial point at a high-amplitude value, say 120, and then look for connected points with amplitudes above a threshold of, say, 100. The software starts by looking at the nearest neighbours of the initial point in all directions: up, down, and in the four orthogonal sideways directions. Voxels with amplitude above the threshold are connected into the body. The process then continues, with nearest neighbours of these new additions being examined to see if they are above the threshold and combining them into the body if they are. The process continues until a picked body is created that in all directions is surrounded by values below the threshold, or which in some directions reaches the edge of the survey volume. If, for example, a hydrocarbon reservoir is marked by bright amplitudes, this is a way of automatically determining the connected pay volume. In practice, trial and error is needed in setting the threshold value. If it is too demanding, the growth of the picked volume will be quickly arrested; if it is too lax, then the connected

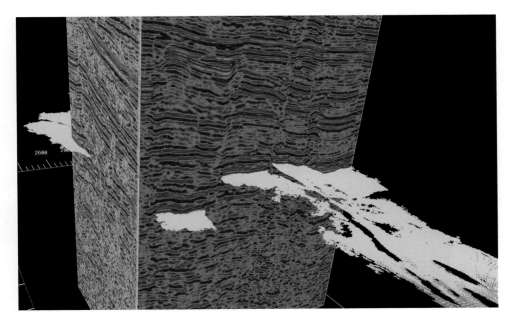

Fig. 7.4 Automatic tracking of bright amplitudes. The 3-D volume can be used to quickly display all data points connected to a given seed point with amplitudes greater than (or less than) a given threshold. This may give a very quick way of sculpting out the reservoir or estimating volumes when used with acoustic impedance data.

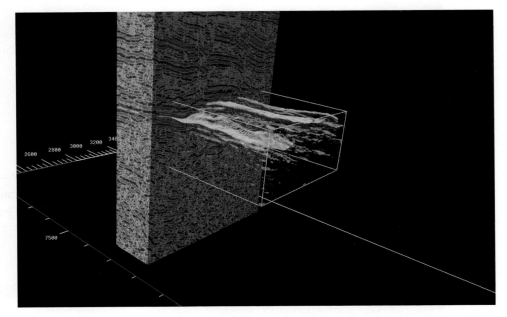

Fig. 7.5 Viewing data in transparency mode. The 3-D data volume can be viewed with part of the data rendered transparent. This allows the interpreter to get a 3-D perspective on the entire reservoir interval, examining issues such as connectivity and rock volumes.

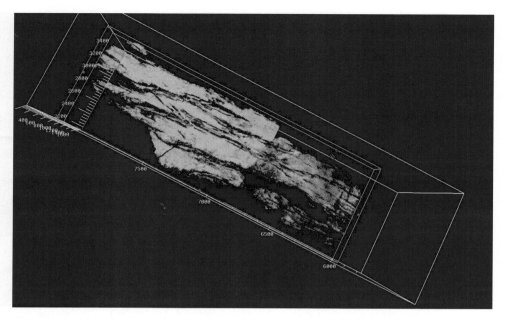

Fig. 7.6 Viewing data from any angle. Here we view from above a 3-D data volume that has weaker amplitudes rendered transparent, giving a good impression of reservoir distribution.

Fig. 7.7 Special glasses worn for 3-D viewing. Polarised glasses are worn that allow the viewer to see the image as a 3-D volume.

Fig. 7.8 Example of full immersive visualisation. New technology includes fully immersive displays where the data are projected onto each wall and the floors and ceilings of a specially designed room. This allows the interpreter to walk through the data, examining details of the reservoir from all angles.

volume will increase to fill much of the 3-D dataset. If the picking is successful, then a 3-D volume corresponding to the 'seismic pay' has been created. This may or may not correspond to the real pay volume. In particular, the limited resolution of seismic means that seismic connectivity is not the same as connectivity between sands in the real earth. Production behaviour or prospect validity may be strongly affected by sands too thin to image seismically.

Figure 7.5 shows another useful display. Part of the data is rendered transparent, leaving only those amplitudes that might correspond to reservoir. For example, all amplitudes less than 100 could be removed completely, and amplitudes in the range 100–128 could be made progressively more opaque. This gives an efficient means of viewing and understanding the 3-D geometry of a reservoir body. This is readily grasped if the semi-transparent cube can be rotated in real time, so that it can be inspected from a variety of angles in quick succession. Setting up the transparency function can be tricky; the idea is to make the voxels of interest bright while removing (making transparent) the rest. Trial and error are inevitable, but a study of well impedances and calculated

Fig. 7.9 Cinema-style set-up of a visualisation centre. Cinema-style rooms with large screens and enough space for teams to view data together have made a large impact on sharing ideas across disciplines.

reflectivity will help in deciding semi-quantitatively what range of voxel amplitudes represents the 3-D body we are interested in.

An advantage of loading the entire 3-D volume is that one can rapidly scan through the complete dataset and view it from any angle. Figure 7.6 shows a display where the data are viewed from above; weak amplitudes have been rendered transparent to make a map view of the reservoir distribution. Techniques such as these have dramatically reduced the time it takes to perform a first-pass interpretation of new exploration areas and have also increased understanding of existing fields.

An advance that is still experimental rather than common practice is the ability to project the data on to a flat screen so that it appears to the viewer as though he is looking at a real 3-D image. The technology is very similar to that used for some films that are viewed in 3-D. The image is displayed as a stereo pair with rapid alternation between the left-eye and right-eye pictures. The viewer wears special glasses that incorporate an electronic shutter so that the left side is transparent when the left-eye picture is on the screen, and the right side for the right-eye picture. Thus the left eye sees only the left-eye picture, and the right eye the right one. In this way, the 3-D perspective is re-created (fig. 7.7). An even more experimental technology is one where the data are projected on the walls and floor of specially built 'visionariums'. These display the data in full 3-D so that the interpreters can even walk through the data volume (fig. 7.8). This technology is in its infancy but may hold potential for revolutionising interpretation in the future (Stark *et al.*, 2000). At present, however, it is rather expensive to implement.

So far we have discussed how working and displaying 3-D volumes, instead of the traditional 2-D slices, has improved the efficiency of the interpreter. However, similar technology has been at the forefront of increased integration across teams and disciplines. Data volumes can be displayed in cinema-style surroundings, allowing many people to view and comment on the data simultaneously (fig. 7.9). Each person brings his own expertise, concerns and solutions, allowing for more informed decisions to be made and ensuring rapid resolution of outstanding issues. It is also an ideal setting for reviewing prospects with senior management since it allows for rapid interaction between the audience and the data.

Reference

Stark, T. J., Dorn, G. A. & Cole, M. J. (2000). ARCO and immersive environments: the first two generations. *The Leading Edge*, **19**, 526–32.

8 Time-lapse seismic

We saw in chapter 5 how seismic data can be used in favourable cases to infer the nature of fluid fill (gas, oil or brine) in a reservoir. An application of this is to follow the way that fluids move through a reservoir during production, by carrying out a baseline seismic survey before production begins and then repeat surveys over the production lifetime. Where 3-D surveys are repeated in this way, they are often referred to as *4-D seismic*, with the idea that time is the extra dimension over standard 3-D. Differences in seismic amplitudes or travel-times between the surveys can reveal the movement of fluid contacts (e.g. where produced oil has been replaced by brine) or the extent of pressure changes that affect reservoir properties. As sketched in figs. 8.1 and 8.2, it is not always straightforward to separate out the causes of the changes. At a producing well, the pressure drops, and if it drops far enough then gas will come out of solution. This will result in a decrease in both P velocity and in density, resulting in a drop in acoustic impedance. On the other hand, the drop in pore pressure causes an increase in P velocity and density. At a water injector (fig. 8.2), we replace oil by water, which in itself would increase P velocity and density; however, the injection also increases the pore pressure, which causes the reverse effect on P velocity and density. By comparing a pre-production trace volume with a post-production volume, we can at least see where the changes are happening, even if we do not completely understand them initially. Figure 8.3 shows three snapshot sections of the reservoir sand in a UK Palaeocene field. We see an initial brightening in the upper sand owing to gas coming out of solution, followed by dimming which may be caused by gas being re-dissolved or by water encroachment. This type of information can be used as a tool for reservoir management, for example by monitoring the expansion of a gas cap, with the intention of managing production so as to prevent it from reaching producing wells and thus decreasing the oil production rate.

Reservoir models are built using well and seismic data. The well data have high vertical resolution, but in many cases there are a limited number of wells available, especially in a high drilling-cost environment such as deep-water offshore. Interpolation between the wells depends on seismic data, which have limited vertical resolution even after attempts have been made to improve it using the methods described in chapter 6. The result is that there may be considerable uncertainty over the possible barriers to

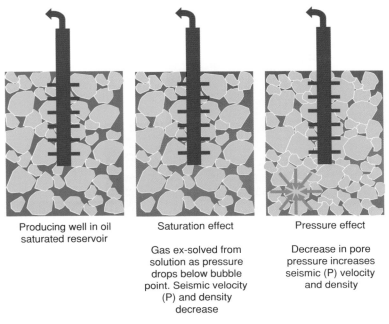

Producing well in oil saturated reservoir

Saturation effect

Gas ex-solved from solution as pressure drops below bubble point. Seismic velocity (P) and density decrease

Pressure effect

Decrease in pore pressure increases seismic (P) velocity and density

Fig. 8.1 Rock property changes around a producing well.

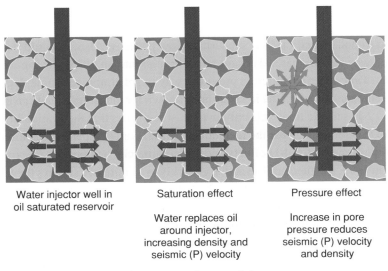

Water injector well in oil saturated reservoir

Saturation effect

Water replaces oil around injector, increasing density and seismic (P) velocity

Pressure effect

Increase in pore pressure reduces seismic (P) velocity and density

Fig. 8.2 Rock property changes around a water injector.

flow, both vertically (e.g. shale bands too thin to resolve on seismic) and horizontally (e.g. faults that may be visible on seismic but whose sealing capacity may not be easy to estimate). Although some understanding of these barriers can be obtained from the observed fluid flow in producing wells, in many cases this will not give sufficient lateral

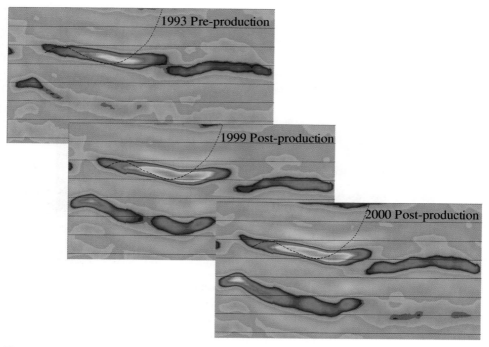

Fig. 8.3   Time-lapse sections. Note the initial brightening in the upper level, followed by slight dimming.

resolution to understand the 3-D subsurface flow patterns. Also, it would be useful to have some way to anticipate gas or oil breakthrough into an oil-producing well before it happens. Time-lapse seismic can provide a way to get almost real-time data on these fluid movements. Although the idea has been known for many years and can be applied to isolated 2-D seismic lines, it is only in the last few years that the availability of high-quality 3-D seismic has made the technique routine. For example, BP acquired its first 3-D survey primarily for 4-D effects and for commercial as opposed to experimental purposes in 1999, and then built up activity quickly to acquire five surveys for 4-D effects in 2001 in the North Sea alone (Ritchie *et al.*, 2002). Shell had acquired 25 4-D surveys worldwide by the end of 2001, and acquired a further 25 in 2002 (de Waal & Calvert, 2002). Like other users, Shell regards time-lapse as proven technology for monitoring fluid movements in thick clastic reservoirs offshore. The challenge is to extend the method to pressure monitoring, to carbonate and to thin-bedded clastic reservoirs, and to the onshore (where significant seismic data quality issues arise).

There are several strands to be combined in any time-lapse study.

- The rock physics that predicts what amplitude and travel-time effects will be produced by changes in fluid fill and reservoir pressure.
- Seismic acquisition and processing issues: we need to distinguish genuine changes in the reservoir from differences in the seismic data due to differences in acquisition,

so to what extent are surveys repeatable? What can be done to match one survey to another if acquisition parameters are not precisely identical?

- Field management issues: when do the time-lapse data need to be available to have an impact on real-world decisions such as where to place an infill well?

An excellent survey of the topic has been given by Jack (1997).

8.1   Rock physics

Modelling of the effect on rock properties of changes in fluid fill, pressure and temperature is an important first step. The issues have been well summarised by Wang (1997). In general, time-lapse is most likely to work when there are large changes in fluid compressibility with production, and in rocks with low elastic moduli (poorly consolidated or with open fractures). For example, time-lapse has been very effective in following steam injection for enhancing recovery of high-viscosity oil from high-porosity sands (Jenkins *et al.*, 1997). In other cases, it may not be certain whether a change in seismic response is great enough to be measured until after a time-lapse survey has been carried out. Effects of changes in fluid saturation are modelled using the Gassmann methodology explained in chapter 5. An empirical observation is that fluid effects are often larger than the Gassmann theory would predict. This may be because clay content is often not sufficiently taken into account when assessing the properties of the rock frame, or because hydrocarbon saturation is patchy rather than uniform. For example, injected water will move preferentially along high-permeability layers within a reservoir and produce a patchy mixture of hydrocarbon and water at the seismic scale.

Understanding the effects of pressure change is more difficult. If the reservoir pressure is allowed to drop, then the net overburden pressure on the reservoir rock increases. For unconsolidated sands or some high-porosity limestones, this may lead to a large reduction in pore volume and thus a significant increase in impedance. For well-cemented sands, the pressure effect will be small. To estimate the likely effect in any particular case, the main source of information is laboratory measurement of rock velocity under controlled pressure conditions. Most laboratory measurements are made at ultrasonic frequencies (around 1 MHz) so that results can be obtained from small samples. Unfortunately, the Gassmann theory is limited to low frequencies; real rocks show considerable change of velocity with frequency (usually referred to as *dispersion*) between the seismic and ultrasonic cases. At high frequencies, there is not enough time for pressure to equilibrate between pores as the seismic wave passes, and rocks appear stiffer than at low frequencies. A method of predicting the size of this effect by modifying the dry rock moduli has been described by Mavko *et al.* (1998). An additional complication of laboratory measurements on core samples is that they may not replicate *in situ* behaviour, mainly because clay minerals can be modified by interaction with pore fluids or by drying.

During conventional oil production, temperature changes in the reservoir are usually fairly small. Injection of cold water (sometimes undertaken to maintain pressure and improve sweep efficiency) may cause detectable changes. Where steam is injected into a reservoir to improve recovery of viscous oils the temperature changes can be large, particularly around the injector wells.

By combining the rock physics data with the output from a reservoir simulator, it is possible to generate synthetic seismic models that show the changes to be expected in the seismic response over time. This is important for planning a time-lapse survey. Knowing the size of the changes will indicate whether time-lapse is feasible at all. A dialogue is also needed with the reservoir engineers. To be useful, a time-lapse survey needs to be available at a moment when it can influence reservoir management, e.g. in deciding the location of an infill well. However, the size of time-lapse effects depends on how much production has taken place. It may not be feasible to detect a time-lapse signal during a very early stage of production, because there has not been enough fluid movement or pressure change to be visible. Choosing when to shoot a time-lapse survey requires collaboration between geophysicists and reservoir engineers. When the survey has been acquired and interpreted, it is usual for the results to be different from the synthetic seismic prediction derived from the reservoir model. The model then needs to be modified so as to give a synthetic seismic response that matches the time-lapse survey. Given the uncertainties in the model and in the seismic data, a range of modifications may match the observed results. Again, dialogue is needed to arrive at a result that can be used to guide reservoir interventions such as modifying oil offtake or water injection patterns.

8.2   Seismic measurements

If we have a baseline and a repeat survey, what can we actually measure that will tell us about the differences between them? There are several possibilities, as follow. In any particular case it will be necessary to calculate the expected change for a given reservoir, fluid fill and pressure change to see how big the seismic changes might be.

(a) TWT to a reflector. Above the reservoir, there should ideally be no TWT differences between the baseline and repeat surveys. In the reservoir itself, seismic velocities will change with pressure or changing fluid fill. The TWT thickness of the reservoir will thus change with production. Seismic events below the reservoir will therefore change in TWT between the two surveys. This effect can be quite large, ranging from a few up to about 10 ms for a thick reservoir interval. In principle such a time shift is easily detectable. Where seismic data have reasonably good signal to noise ratio, it should be possible to pick a reflector to an accuracy of 1 or 2 ms. Where we are dealing with a shift between one suite of reflectors (above the reservoir) and another (below the reservoir), then correlation methods should give even

better accuracy. Such time shifts are often the time-lapse effect that is most robust against differences in acquisition and processing between baseline and repeat
surveys.

(b) Amplitudes of reflectors. As explained in chapter 5, the amplitude of a reflector (e.g.
at top reservoir) may be diagnostic of its fluid fill. Changes in amplitude between
baseline and repeat survey may therefore be diagnostic of changes in fluid fill. As
we shall see, amplitude changes between the surveys may also be caused by several
factors related to data acquisition and processing, but it can still be possible to
extract useful time-lapse information.

(c) Seismic velocities. Subsurface velocities are routinely estimated as a step in stacking
and migrating seismic data, and in principle they will change with change in fluid fill.
However, the accuracy with which stacking velocities can be estimated is limited;
it will often be hard to get a higher accuracy than 1%. In a realistic case, the
effect of a change in fluid fill on stack or migration velocity to a base-reservoir
reflector will usually be less than this. The problem is that the velocity change
is confined to a fairly thin layer, but stacking velocities are an average velocity
from surface to the reflector concerned, and are dominated by the thick unchanged
overburden. Also, as we saw in chapter 3, stacking velocity may change with
acquisition azimuth owing to genuine anisotropy or to dip and curvature of the
reflector; it will also change with azimuth and maximum offset due to near-surface
effects. It is therefore sensitive to acquisition differences, particularly when very
high accuracy is wanted. All this makes velocities hard to use as a time-lapse
diagnostic.

If we are intending to make use of differences in TWT and amplitude between
surveys, there are two possible approaches. Where fluid or pressure effects are large,
and fluid movement follows a simple pattern, it may be sufficient to make maps of TWT
and amplitude for a reservoir-related reflector on both surveys, and compare them. If,
for example, hydrocarbons cause obvious bright spots at top reservoir, then the extent
of the bright spots on amplitude maps of the two surveys can be compared visually.
This does not require that the acquisition or processing should be exactly the same
in the two surveys. It should certainly follow good practice for obtaining reasonably
accurate amplitudes, but what we are looking for is a lateral change in amplitude on
each map independently (marking the fluid contact), followed by a comparison of the
positions of these amplitude changes on the maps of the two surveys. If, on the other
hand, we are looking for small effects, then we may have to subtract one trace volume
from the other in order to reveal subtle differences, such as small amplitude changes
within a reservoir interval caused by interfingering of oil and water in the producing
zone. This is much more difficult to do, because there are many possible reasons for
differences between surveys that are related to acquisition and processing differences.
These differences will appear as unwanted noise in the difference volume, and their
elimination requires careful processing.

8.3   Seismic repeatability

If we are thinking of comparing travel-times, amplitudes or seismic velocities between a baseline and a repeat survey, there are several possible other causes of differences between the surveys that will interfere with our comparison:

(a) Noise. Ambient noise can easily vary from one survey to another. For example, if the repeat survey is being carried out over a producing field, there will probably be many sources of industrial noise that were absent when the pre-production survey was shot. This need not be a significant problem so long as the signal to noise ratio at reservoir level remains high.

(b) Accessibility. A common problem is that some areas that were shot in the baseline survey are not accessible to the repeat survey because of production facilities. For example, the presence of a production platform offshore will create a hole in the survey acquisition because it will not be possible for a survey vessel to approach it very closely for safety reasons. This means that over part of the area of interest, it will not be possible to acquire a survey that closely replicates the original acquisition pattern.

(c) Near-surface effects. There may be quite large changes in the near-surface over time, for example as a result of changes in the depth to the water-table with the season of the year. Although in principle it is possible to allow for these effects, they introduce an additional uncertainty when comparing two surveys. The shifts may be comparable to or greater than the expected time-lapse time shifts, but of course the latter would be confined to the interval below the reservoir whereas near-surface effects will cause a shift of the entire trace. In the marine case, similar effects can be caused by a change in the sound velocity in the sea-water column, owing to temperature changes caused by large-scale currents. Tidal effects also need to be allowed for.

(d) Source signature. This may differ markedly from one survey to another. Ideally, seismic data should have been converted to zero-phase during processing (chapter 2). If this has been successful, it will remove much of this problem.

(e) Acquisition parameters. There may be differences in the geometry of the source and receiver pattern between the surveys. Jack (1997) quotes some examples of effects that were simulated by taking a well-sampled survey and removing some of the data, followed by processing of the decimated dataset. For example, when an 80-fold marine dataset was decimated to 40-fold, then the difference section between the final 80- and 40-fold sections along a typical dip line had an rms amplitude 40% of the original section. In such a test, because the decimated data were produced by merely dropping traces from a 'baseline' survey, positioning repeatability is not an issue. In real data, two problems arise. One is that, even if we knew the positions of sources and receivers precisely for the baseline survey, it might not be possible to replicate them. This is typically the case for a marine survey, where streamer locations are affected by surface currents; recent technology allows steering of the individual streamers, which will reduce, but not eliminate, the problem. Secondly, there will always be inaccuracies in position-fixing of shots and receivers. Considerable improvement in positional accuracy of marine surveys has been achieved from about 1990 onwards, owing to satellite positioning, the tracking of streamer tailbuoys, and the use of acoustic transponder networks. If the baseline survey is old enough to have been shot before these improvements, then there may be quite large uncertainties on the exact locations of sources and receivers. One simulation showed that errors of 80 m in the location of the tail of a 4000 m marine streamer could give rise to difference sections with an rms amplitude 20–30% of the baseline data. Study of VSP data by Landro (1999) showed that shifting a shot location by 10 m resulted in an rms change in the record (after static alignment) of 20%, and a 20 m shift resulted in a 30% change. The records were dominated by the downgoing signal, so these changes are caused by transmission response variations between neighbouring ray-paths. One way of reducing acquisition differences is to leave receivers permanently in place on the seabed, as has been tried in the BP/Shell Foinaven Field. Even so, difference sections between surveys showed rms amplitude differences up to 35% of the original data (Jack, 1997).

(f) Processing parameters. There are many steps in the processing sequence that can introduce differences similar to the time-lapse signal we are looking for. These include statics correction, mute design, pre-stack deconvolution, stack and migration velocity derivation, and amplitude balancing. It is often necessary to reprocess the baseline survey together with the repeat survey, as discussed in the next section.

8.4   Seismic processing

Specialised processing can alleviate these problems of repeatability. Where the baseline survey is a 'legacy' dataset, not originally acquired with time-lapse in mind, it will be beneficial to reprocess it in parallel with the monitor survey. This ensures that the same algorithms are used at each stage, with so far as possible the same choice of parameters. At each step, the surveys can be compared and action taken to match them if necessary. This might include time- and space-variant amplitude and spectral trace matching between the two datasets. A detailed discussion is given by Ross *et al.* (1996). The idea is to match the two surveys over the intervals outside the reservoir, where no time-lapse changes are expected. The quality of the match can be seen from the difference volume between the baseline and monitor surveys, which should ideally be zero except where the production-related effects are present. The matching process can include several elements. Time corrections are needed where there are systematic shifts, for example as a result of salinity variations in the water column in the marine case, or changes in the near-surface velocities in the land case. Timing differences of as little as 2 ms can give significant residual energy in the difference plot. Root-mean-square energy balancing using a smoothed time and space variant scalar is usually required, even if the baseline and monitor survey were nominally acquired in an exactly identical way. Spectral matching between the two surveys will also, in general, be required to ensure that they both have the same frequency content. Finally, differences in seismic phase must be removed. A phase difference of as little as 15° will be significant for the difference plot; such small phase shifts are typically too small to be apparent to the unaided eye, and rely on software for their detection and correction.

8.5   Examples

Two interesting examples were presented by Hatchell *et al.* (2002), who discussed the process of comparing time-lapse results with predictions from a reservoir simulator. Figure 8.4 shows the time-lapse difference for both real and synthetic data over the Schiehallion oil field in the West Shetland area. The main sources of time-lapse response are pressure and saturation effects near the wells. Near the water injectors, reservoir pressures have increased from 17 MPa to 41 MPa, which causes brightening on the seismic. Near producing wells, the pressure drop causes an amplitude decrease as long as the pressure is still high enough for gas to remain in solution; once the pressure is low enough for free gas to be present, the amplitude brightens. All these effects were visible on both the real and the synthetic data, and detailed examination of the real data allowed the mapping of pressure compartments separated by sealing faults, some of which are too small to be resolved by conventional imaging. Another example is from the Maui gas field in New Zealand. The reservoir interval contains laminated sands and shales,

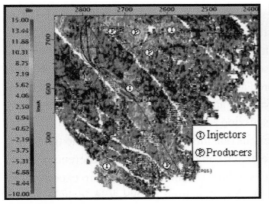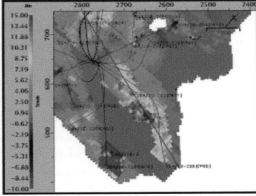

Fig. 8.4   Schiehallion field: difference in amplitudes between pre- and post-production for both real (left) and synthetic (right) seismic. Reproduced with permission from Hatchell *et al.* (2002).

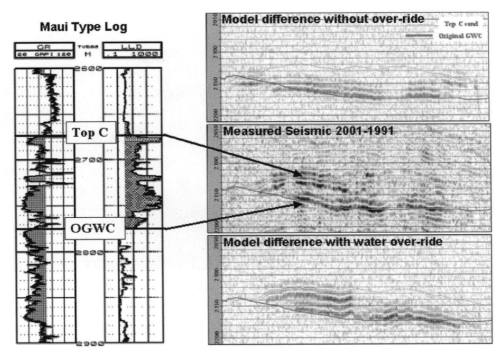

Fig. 8.5   Maui field: type log across the reservoir interval, and comparison of models to real time-lapse seismic. Reproduced with permission from Hatchell *et al.* (2002).

and the question was whether water is simply displacing gas vertically throughout the reservoir or whether the water is preferentially moving through the uppermost and most permeable sands, over-riding unproduced gas in the less permeable sands below. Figure 8.5 shows a comparison between the real time-lapse data and synthetic models for the two cases. Clearly, the real data are in close agreement to the prediction with water over-ride.

Stammeijer *et al.* (2002) have published the example shown in fig. 8.6. This shows time-lapse results over a mature field in the Northern North Sea that is being produced by water-flood. The baseline survey was acquired in 1995, some six years after production began, and the repeat survey was acquired in 2000. In this case, connected volume analysis (as discussed in chapter 7) was used to identify the zones where oil had been swept from the reservoir in the period between the two surveys. Some of the zones where no change has occurred can be identified as areas completely swept before 1995. The others represent unswept areas, which can now be targeted by an infill drilling programme.

Repeat seismic surveys over a field in the UK West Shetland area allowed a flood front to be seen (fig. 8.7). Shortly after the repeat seismic survey showed that the flood front was approaching the producer, significant water production began to be observed.

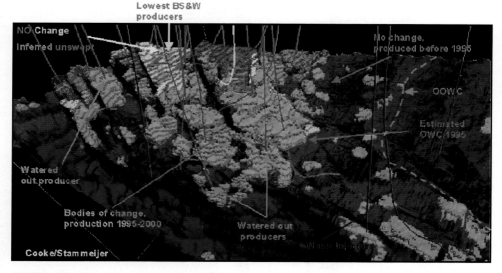

Fig. 8.6   Northern North Sea field. Perspective display of field structure, with wells and reservoir bodies indicated. Blue bodies: oil swept between 1995 and 2000. Reproduced with permission from Stammeijer *et al.* (2002).

Flood Front Monitoring

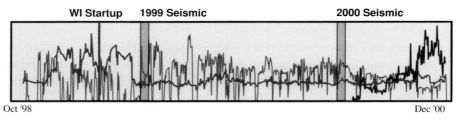

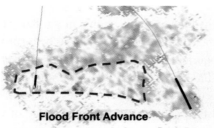

Flood Front Advance   2000 – 1993: Flood Front Visible

Fig. 8.7   In the upper graph, green is oil production rate, blue is water production rate and red is the gas–oil ratio of the produced fluid. The difference map of amplitudes between the 1999 survey and the 1993 pre-production baseline (not shown) revealed brightening due to gas coming out of solution. The difference map between 2000 and 1993 shows brightening (green) in the peripheral areas owing to gas evolution, but dimming (blue) around the water injector. The blue zone therefore shows the position of the flood front in 2000, and is consistent with the subsequent water breakthrough at the producing well.

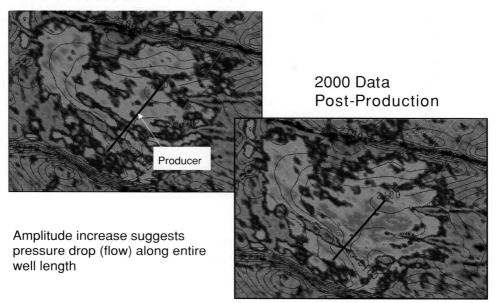

**1999 Data
Pre-Production**

**2000 Data
Post-Production**

Producer

Amplitude increase suggests
pressure drop (flow) along entire
well length

Fig. 8.8 Amplitude increase (green) suggests that pressure drop is occurring everywhere in the vicinity of the producer well; there had been concern that it was producing only from some segments of the faulted and channelised reservoir. A 4-D seismic survey was cheaper than establishing the flow from each section of the well by direct measurement, which would have interfered with production.

In another field in the same area, repeat surveys established that a well was effectively draining the whole area around it (fig. 8.8).

References

de Waal, J. A. & Calvert, R. (2002). 4D seismic all the way – implementing time lapse reservoir monitoring globally. *Abstract H-01, EAGE 64th Annual Meeting, Florence.*

Hatchell, P., Kelly, S., Muerz, M., Jones, C., Engbers, P., van der Veeken, J. & Staples, R. (2002). Comparing time-lapse seismic and reservoir model predictions in producing oil and gas fields. *Abstract A-24, EAGE 64th Annual Meeting, Florence.*

Jack, I. (1997). *Time-lapse Seismic in Reservoir Management.* Distinguished Instructor Short Course, Society of Exploration Geophysicists.

Jenkins, S. D., Waite, M. W. & Bee, M. F. (1997). Time lapse monitoring of the Duri steamflood: a pilot and case study. *The Leading Edge*, **16**, 1267–74.

Landro, M. (1999). Repeatability issues of 3-D VSP data. *Geophysics*, **66**, 1673–9.

Mavko, G., Mukerji, T. & Dvorkin, J. (1998). *The Handbook of Rock Physics.* Cambridge University Press.

Ritchie, B., MacGregor, A., Strudley, A. & Goto, R. (2002). The impact of new 4D seismic technology on the Magnus Field. *Abstract A-20, EAGE 64th Annual Meeting, Florence.*

Ross, C. P., Cunningham, G. B. & Weber, D. P. (1996). Inside the crossequalisation black box. *The Leading Edge,* **15**, 1233–40.

Stammeijer, J., Cooke, G. & Kloosterman, H. J. (2002). 4-D driven asset optimisation. *Abstract A-26, EAGE 64th Annual Meeting, Florence.*

Wang, Z. (1997). Feasibility of time-lapse seismic reservoir monitoring: the physical basis. *The Leading Edge,* **16**, 1327–9.

Appendix 1

Workstation issues

This appendix covers briefly the main issues involved in managing hardware, software and data to create an environment for the interpretation of 3-D seismic data. No attempt will be made to discuss specific details of individual vendors' offerings, as they change very quickly; rather, the objective is to give a general overview of the requirements for the creation of a successful interpretation environment.

A1.1 Hardware

The volume of seismic data in a 3-D survey is often large. As we saw in chapter 3, a modest survey may contain 500 000 traces each of 1000 samples. A large survey might contain tens of millions of traces. It is quite usual to have several versions of the data volume (e.g. near and far trace stacks for AVO analysis, perhaps several different inversion results, and calculated attribute volumes such as coherence). Storage requirements for the trace data may amount to several to a few tens of gigabyte (1Gbyte = 10^9 byte). If a company has interests in a number of licences, each with its own 3-D survey, the total volume of seismic data may amount to a few terabyte (1 Tbyte = 10^{12} byte). Storing such a volume on disk is possible, but the cost will be less if some of the data are held offline on magnetic tape, preferably in a high-density format (e.g. helical scan) that will allow large volumes (e.g. 100 Gbyte) to be stored on a single tape. Rapidly changing commercial priorities will dictate that the archive of data on tape is not static, but will need to be retrieved and reworked, maybe with only a day or two's notice. Robust data administration procedures are needed to make this possible.

A further complication arises in a large integrated project, where several users may need access to the same seismic dataset. Instead of each interpreter having his or her own workstation which runs interpretation software on data held in local disk storage, issues of cost and version control may then dictate that all data are held centrally and accessed by software that may also be run on the central server or may be run on the interpreter's workstation. Computer power and data access time on the central server and bandwidth of the network between it and the interpreter's seat will then dictate system performance, and need careful consideration to get good performance in a large installation.

As with all computer systems, backup of work in progress is critical. The interpreter's work needs to be backed up onto tape, preferably at the end of every working day. Many installations have found that proper backup is more certain to be carried out if it is managed centrally rather than by the individual interpreter. This is another driver towards centrally held and managed data stores. Where there are many interpreters involved, the amount of data to be backed up each day will become large enough to need detailed planning; high-density tape drives are a suitable medium, but an efficient system is needed to make sure that all the required backup is completed in the time available.

The interpreter spends a large part of his or her working day looking closely at data on a screen display. High-quality display screens are therefore important. As with any work involving sustained use of a screen, mouse and keyboard, the physical design and siting of the units needs careful consideration to avoid possible damage to health.

A1.2   Software

Decisions on which software should be made available on the system depend on consideration of functionality and cost, and need to be assessed in detail for each individual installation. The main problem from the point of view of the system administrator is usually the need to provide data transfer between applications from different vendors. This can of course be reduced by restricting software purchases to a single vendor, but even then there is likely to be a need to interface with proprietary applications and databases. Where applications from several vendors are in use, system upgrades (e.g. major updates to operating systems) can be very arduous, requiring updates to both software and databases. Particularly when there is doubt about the reliability of hardware or software in the new environment, it may be necessary to run both old and new environments in parallel, with a phased transfer from one to the other; this needs careful liaison with interpreters and other users of seismic data to avoid service interruptions at moments that are critically important to the business.

A1.3   Data management

Management of seismic trace data is for the most part not difficult, except for the huge volumes involved. The data are organised in a very regular fashion, as values of amplitude (or other attribute) on a regular 3-D grid. When data are retrieved, it is almost always along a plane or line or sub-cube of the 3-D grid (e.g. as sections or traces). There is no requirement to identify individual samples that satisfy complex selection criteria. This means that the data do not have to be held in a complex database structure. Once the data have been loaded, usually from SEGY tapes, the files require little further maintenance. Data loading is usually fairly easy for 3-D seismic traces. The main problem is the specification of the spatial location of the traces, which is often done by manually typing in real-world co-ordinates of particular points, e.g. of three corner points of the rectangular survey grid. This process is potentially error-prone; small mistakes are difficult to spot once the data have been loaded but could have disastrous consequences for accurate well positioning. Furthermore, there is scope for confusion between different co-ordinate systems (projection, spheroid and datum). Sometimes position data are supplied on one projection system and need to be transformed to another, to match against corporate standards or to merge with another dataset. It is useful to have the services of a topographic surveyor to ensure that the co-ordinate data as loaded are correct and in the desired projection system.

Loading seismic traces for a set of 2-D lines is much more difficult. The main problem is to work out the relationship between trace number on tape and position data on a map; this can be quite simple where the traces and positions are numbered sequentially along a line, but where a line has been shot in several parts with discontinuities in numbering of traces and surface locations, it can be very time-consuming to work out the relationships unless there is excellent paper documentation.

Management of seismic interpretation data (i.e. picked horizons and faults) is less simple. Because it is easy to create new horizons, an interpreter at the end of a mapping project may have defined hundreds of them. Most of them will be trial efforts to investigate different features of the data, perhaps over very limited areas; usually a few dozen horizons will contain all the significant information. Cleaning up interpretation projects to throw away all the useless horizons should be part of the reporting process at the end of a project; if this is not done, it is very difficult for another interpreter to come to grips with the interpretation and to use it as a basis for further work. To make such re-use easier, there also need to be fairly rigid naming conventions for horizons, linked to the stratigraphic column. Over a period of time, a given seismic volume will be interpreted by a number of people, and different versions may exist of the same stratigraphic horizon (depending, for example, on the view taken about correlating horizons into undrilled fault blocks). In addition, a given geographical area may be covered by several different interpretation projects, corresponding to different seismic surveys or different reprocessed versions of a single survey. The interpreter may need to review all of them, and carry forward several of them into the next stage of study (e.g. reservoir modelling), in order fully to reflect the degree of uncertainty about the subsurface (Herron, 2001). It may be hard to find out what interpretations are available, let alone deciding which is the best for any particular purpose. Ideally, a database is required showing available interpretation data by geographic area, stratigraphic level, etc.; it is not difficult to devise a simple scheme, but maintaining such a database will have a significant manpower cost.

Finally, the interpreter needs to be able to access well data to tie the wells to the seismic survey. Sonic and density logs, at least, need to be available in digital form, so that well synthetics can be created. To understand seismic response and its lateral variation, it is helpful to have other wireline logs as well; for example, the caliper log, which measures borehole diameter, is a useful indicator of log reliability, because where the borehole diameter is locally much larger than expected the logging tools may not be able to record correct formation parameters. The interpreter also needs stratigraphic and lithological information. In general, management of well data is not easy. Stratigraphic information in particular can be complicated, with repeated or missing sequences in any particular well and lateral changes in nomenclature. A system designed primarily for use by geologists may be over-complicated for the needs of the seismic interpreter. For the creation of well synthetics, it may be useful to set up a separate database (or partition of a larger database) containing sonic and density curves specially edited for synthetic creation, together with checkshot information, major formation tops and perhaps the caliper log and the gamma-ray log for correlation with other displays and for identification of sand/shale trends. The issue here is that, as we saw in chapter 3, the well synthetic can be sensitive to noise in the sonic and density logs. Where wireline logs are being used primarily for formation evaluation, defective readings in the shales will probably not get paid much attention; therefore, there is a need to edit logs specifically for synthetic generation, and once edited they might as well be retained in a simple database structure.

Reference

Herron, D. A. (2001). Problems with too much data. *The Leading Edge*, **20**, 1124–6.

Appendix 2

Glossary

Accommodation space

The space available for potential sediment accumulation, controlled by processes such as changes in sea-level, tectonic movements, compaction of pre-existing sediment and subsidence.

Acoustic impedance

A property of a rock, defined as the product of density and seismic velocity. The impedances on either side of an interface determine the reflection coefficient for seismic waves (section 3.1).

Aeolian

An aeolian depositional process is one where the dominant agent is wind. A typical modern example is a hot arid desert such as the Sahara. There may be areas covered with sand dunes, and other areas of bare exposed bedrock. Large areas of sand dunes form sand seas or *ergs*. Aeolian sandstones can be important hydrocarbon reservoirs, as for example in the Permian of the North Sea.

AGC

Automatic Gain Control is the process of varying the gain of a seismic trace display with TWT, so as to maintain the average absolute level constant within a time window. If a short time window is used, the process has the effect of destroying information about lateral and vertical changes in reflection strength, which is highly undesirable if any attempt is being made to recognise effects due to fluid content or lateral change in lithology. If a long window (e.g. 1 s) is used, however, the process makes it easier to view seismic displays across the full TWT range, and can even be beneficial to amplitude studies by removing effects of variation in near-surface absorption, for example.

Aliasing

Seismic traces are not continuous measurements in time; instead, the amplitude is recorded digitally at a certain sample interval in time (usually 2 or 4 ms). Nor do we record traces at every possible (x, y) location; rather, we shall have traces on a regular grid, perhaps 25 m × 25 m. All our data processing and interpretation assumes that this is an adequate approach, in the sense that a smooth curve through the sampled points is a true picture of reality. This would obviously not be true if there were strong high-frequency variations present in the data, so that in reality there were high-amplitude oscillations happening between the sample points. It can be shown that signals can be recovered from a regularly sampled representation provided that they do not contain frequencies higher than half of the sampling frequency. Frequencies higher than this cannot be recovered, and are said to be aliassed; they will mimic the behaviour of a lower-frequency signal. Care is therefore needed not to alias signals when they are recorded; if there is a danger of this, the sampling frequency must be increased or the bandwidth of the incoming signal decreased before it is sampled for turning into digital form.

Alluvial fan

On land, rivers are the main means of transporting sediment. Alluvial fans are localised areas of high sediment accumulation, formed at a place where laterally confined flows are able to expand, for example on leaving a gorge to flow into a broad valley floor. The expansion of the flow leads to a velocity decrease and a reduction in ability to transport sediment, which is therefore deposited.

Amplitude spectrum

The methods of Fourier analysis allow us to envisage a seismic trace as the sum of a series of traces that are pure sine waves. If we perform this analysis and then plot the amplitude of the sine waves against frequency, we have the amplitude spectrum of the original signal. Seismic processing often aims to produce a reasonably flat amplitude spectrum in the frequency range from, say, 8 to 50 Hz, though what is possible depends on the seismic source, the propagation path length and the degree to which the rocks absorb high frequencies.

Angle stack

It is often possible to infer subsurface lithology or fluid fill from the way that reflection amplitude varies with incidence angle on a reflector (AVO). One way to investigate this effect is to make sub-stacks of the data, including only data from a particular angle range. Thus, one might compare a stack containing incidence angles 0–15° with a 20–35° stack. These datasets are produced by stacking the appropriate range of offsets, which will vary with TWT; the required offset range for a particular set of angles at a given TWT can be calculated if the seismic velocity is known as a function of depth, either from well information or seismic velocity analyses.

Anisotropy

In general, anisotropy means that a physical property varies with the direction in which it is measured. In our case, the property of interest is usually seismic velocity. In most cases, this exhibits transverse isotropy; there is a symmetry axis, and velocity is the same in all directions in the plane perpendicular to this axis, though different from the value parallel to the axis. Fine layering (e.g. in a sand–shale sequence) of materials with different velocities will produce rocks with a vertical symmetry axis. Vertical cracks, all oriented in the same plane, would produce a horizontal symmetry axis, perpendicular to the cracks.

Attributes

The first measurements made on seismic traces were of travel-time to a reflector, with a view to making a structural map of it. As data quality improved, it was realised that the amplitude of a reflection could carry useful information; changes in amplitude along the reflector might relate to lithology or reservoir quality, for example. There are many different ways in which we might measure amplitude. The maximum excursion of a loop is the most obvious, and would, for example, be appropriate if we were looking at an isolated reflector at the top of a massive sand overlain by a massive shale. On another occasion, we might be looking at an interbedded sand–shale sequence, where there would be a whole series of reflections, each one may be laterally discontinuous, between the top and the base of the sequence. If we wanted to characterise the variability of the sequence, it would be useful to measure the average amplitude (without regard to sign, so average absolute value or rms average) over the interval between the top and base of the package. Or again, we might also get information from the shape of a seismic loop, for example where we are trying to follow thin beds near the limit of seismic resolution; the most obvious measure might be the width of the loop, i.e. its duration in TWT between zero-crossings, but many other shape measures are possible. 'Attribute' is in use as a loose generic term to cover all these types of measurement on a seismic trace. It usually refers to measurements taken on a limited time-window around a reflector or interval of interest, and very often to measurements made on single traces, though some attributes are measured over a small cluster of traces (e.g. the local dip or curvature of a reflector).

Autotracker

Software to follow a reflector automatically through a seismic trace volume, starting from a limited amount of manually picked data (*seed points/lines*).

AVO

Amplitude Versus Offset: the variation of reflector amplitude with source–receiver separation (see chapter 5). Sometimes denoted AVA (Amplitude Versus (Incidence) Angle).

Avulsion

Overbank sedimentation from an active river channel causes the growth of a ridge above the surrounding floodplain, within which the channel is contained. When a channel bank is breached in a flood, the river can find a new path on a lower part of the floodplain, leading to the abandonment of the old channel. This process is called avulsion, and occurs in delta distributary channels as well as meandering rivers.

Bin

The final processing output of a 3-D survey will be a set of traces on a rectangular grid. It is not feasible to acquire data on a perfectly regular grid. Therefore, a uniform grid of rectangular *bins* is superposed on the acquisition map. Each acquisition trace is assigned to the bin located at the point midway between its source and receiver positions. The traces falling within a bin will be stacked after correction for normal moveout, giving the uniform output grid of traces. Some refinements of this basic idea are discussed in chapter 2.

Bright spot

The classic hydrocarbon indicator is the bright spot, when pay (usually gas-bearing) sands are very soft and brine sands are similar in impedance to the overburden. The top reservoir event will then be a strong soft loop in the pay zone, and a weak event elsewhere. Such an amplitude anomaly will conform to structure, unless there is a strong lateral change in lithology across the area of hydrocarbon fill.

Bulk modulus

If a volume V of a medium is acted on by a pressure P, the volume will be decreased by ΔV. The dilatation Δ is defined as $\Delta = \Delta V / V$, and the bulk modulus k is defined as

$$k = -P/\Delta$$

where the minus sign is inserted to make k positive.

Checkshot survey

In order to tie well information to seismic data, it is important to know the relationship between depth and TWT at the well. This can be established by measuring the travel-time from a source on the surface to a receiver at a known depth in the well. This information is usually collected at a number of receiver depths. In the case of a deviated well, either the source can be located vertically above the receiver (and thus at some distance from the wellhead), or corrections can be applied for the slant travel path.

CDP gather

Synonym for CMP gather (*see* **CMP gather**).

CMP gather

The Common Midpoint gather is a collection of traces having different source–receiver offsets but the same midpoint location between source and receiver. For horizontal reflectors, the traces of the CMP gather have the same reflection point and can therefore be stacked together after NMO correction. In the presence of reflector dip, DMO correction will be needed before stack.

Crevasse splay

Where a river is confined between levees, from time to time during floods the water level will reach the top of the levee and spill over. At a place where this happens, a shallow crevasse is formed on the crest of the levee. From the crevasse, lobes of silty or sandy sediment spread onto the floodplain as crevasse splays. Similar features are formed in the distributary channels of deltas.

Crossline

See **Inline**.

Datum

On land, there may be appreciable topography across a survey area. Furthermore, if explosive charges are used as the source then they may be buried at varying depths across the survey. Also, there will usually be lateral changes in the thickness of the near-surface low-velocity layer. To remove all these complications, a reference datum is selected and corrections are made to travel-times so that they represent what would be recorded if shots and receivers were placed on the datum surface, it being assumed that there is no further low-velocity layer below this surface. For onshore surveys, the datum surface may be flat or it may be a more or less smoothed version of the topography. Offshore, mean sea-level provides a convenient datum.

Decimation

Reduction of the number of traces in a dataset by systematic removal of, for example, every other trace. It is not restricted, as the name suggests it might be, to the removal of every tenth trace.

Deconvolution

The seismic trace can be thought of as the result of convolving the earth reflectivity with a wavelet. Ideally, we would like to have a seismic source which gave out a single sharp spike signal, and record a reflected signal which was simply a series of spikes. Unfortunately, any real source will emit a signal of finite length. Furthermore, the source signal will be modified as it passes through the earth, because of absorption, scattering, and other causes. The result is that our recorded signal will be the sum of the reflections of a wavelet from the series of subsurface reflectors (fig. A2.1). Mathematically, this process is represented by the convolution of the wavelet with the earth reflectivity. Deconvolution is a signal processing step that attempts to undo this convolution, to leave us with the earth reflectivity series, thus improving the resolution of the seismic data.

Delta

A depositional system formed where a river supplies sediment to a coast, forming a shoreline protuberance. The input of mud and sand from the river is reworked by marine processes (wave and tide action). The components of a delta system (from onshore to offshore) are: (1) the delta plain, largely sub-aerial and consisting of distributary channels separated by inter-distributary areas of marsh and swamp; (2) the delta front, where fluvial and marine processes interact and form features such as sand bars at the distributary mouths; (3) the prodelta, a zone of quiet deposition, typically of mud and fine silt, from suspension. As time goes by, if sediment continues to be supplied, the delta builds outward from the coast. In vertical succession, the prodelta muds will then be overlain by progressively shallower-water deposits, generally with higher silt and sand content.

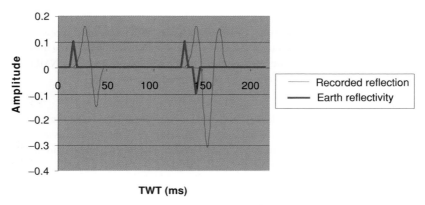

Fig. A2.1   Earth reflectivity, in red, shows that there are three reflectors. The trace in purple shows what would actually be recorded; each reflector sends back a signal representing the wavelet, scaled in amplitude by the reflection strength, and the final output is the sum of these three signal trains. It is hard to visualise the earth reflectivity from the recorded trace when the reflectors are closely spaced. Deconvolution tries to remove this confusion.

Depth migration

An implementation of migration that allows for ray-bending at interfaces where the seismic velocity changes. The simplest type of migration (time migration) can cope with mild lateral changes in velocity by changing the migration velocity (and therefore the curvature of the hyperboloid along which data are summed in the Kirchhoff method). Where there are rapid lateral changes, such as in the presence of steeply dipping interfaces across which there are large velocity changes, then the surface along which data ought to be summed will no longer be a hyperboloid and has to be calculated in detail by tracing the propagation of rays through a subsurface model. This is substantially more demanding in computation time. It is also more difficult to build the correct subsurface velocity model, because changes in velocity at any one level will affect rays traversing that level in ways that may be hard to visualise without detailed computation.

DHI

A Direct Hydrocarbon Indicator is any feature on seismic data that gives evidence for the presence of hydrocarbons, and is particularly useful in reducing the risk associated with drilling an exploration well. Typical examples are an amplitude anomaly on the top-reservoir reflector that shows confor- mance to structure, or a *flat spot*, a horizontal reflector due to a gas–oil or oil–water contact that cuts across bedding-plane reflectors.

Dip moveout (DMO) correction

In the simplest approach to stacking, traces with the same source–receiver midpoint are added together after correction for NMO. The output stacked trace is then treated as being a lower-noise version of the trace that would be recorded if the source and receiver were coincident at the midpoint location used to select the input traces. For reflection from a dipping interface in the subsurface, the reflection point is not vertically below the source–receiver midpoint. It is displaced updip, by an amount which increases as the source–receiver distance increases. Stacking traces with the same midpoint therefore adds together traces that have been reflected from different points in the subsurface, inevitably smearing the image. DMO processing corrects for this effect. It carries out a partial migration of the data, moving energy for a reflecting point from the location seen on an actual finite-offset trace to the location that it would have for a zero-offset trace. This involves shifting the energy both laterally and vertically. After DMO has been applied, traces can be stacked together at each CMP surface location without smearing the image, because all the traces contain reflected energy from the same subsurface point.

Dip section

Seismic section shot parallel to the predominant geological dip direction, and so in the direction of maximum horizontal gradient of reflecting horizons and often perpendicular to the principal faults.

Fold

As can be seen in fig. 2.7, seismic acquisition is normally laid out so that traces recorded at different source–receiver offsets can be added together ('stacked') to enhance signal to noise ratio. (They will, of course, require correction to be made for the more oblique ray-paths at the longer offsets.) The number of individual source–receiver pairs that contribute to the stack is called the fold of the data. Around the edge of a survey area, the fold decreases because there will be progressively fewer long offsets available. The 'full fold' area is therefore surrounded by a zone where the fold progressively tapers to 1. Because of the increased noise level, the data in the zone of reduced fold are of limited use and are often largely excluded from the migration process.

Fresnel zone

If we think in terms of rays, then a reflection comes from a point on a reflecting surface. In terms of wave theory, however, a reflection is made up of energy returning from a finite area of the reflector. A Fresnel zone is the area from which reflected energy arriving at a receiver has a phase difference of no more than half a cycle, and is therefore able to contribute constructively to the reflection. For the case where source and receiver are coincident and a distance h above the reflector, then most of the reflected energy comes from a circular zone of radius r given by $r^2 = \lambda h/2$, where λ is the wavelength of the seismic signal.

GPS positioning

The Global Positioning System depends on time-ranging to a set of satellites, which are distributed in various orbits so that a user can receive signals from at least four of them at any point on the Earth's surface at any time. Since the satellite positions in their orbits are known to high accuracy, four range measurements are enough to calculate the user's latitude, longitude and height, plus the timing offset between the user's clock and the satellite system clock. For the highest accuracy, differential GPS (DGPS) is used, where a fixed station monitors its apparent GPS position and the deviations from its known location are used to refine the apparent position of a user in the field. In this way it is possible to correct for uncertainties in orbital parameters, atmospheric refraction, and deliberate signal degradation by the system's owners, giving a positional accuracy of, for example, 2 m within 2000 km of the reference station.

Inline

A 3-D seismic survey consists of traces on a rectangular grid. One of the axes of this grid is called the inline direction and the other is called the crossline. Lines in these two directions are numbered, and then co-ordinates of traces within the survey can be specified by means of the (inline, crossline) co-ordinate pair. The choice of which direction is called inline and which crossline is sometimes arbitrary. The inline direction may refer to the original shooting direction for the survey, but a bin grid may be used for processing that is not simply aligned to the original acquisition layout.

Invasion effects

When a borehole is drilled, the drilling fluid will penetrate into the formation being drilled. When wireline logs are run to measure formation density and seismic velocity, for example, then they may record not the true formation values, but rather those in the zone around the borehole where the drilling fluid has *invaded* the formation, replacing the original fluid. Particularly where oil and gas reservoirs have been drilled with a water-based mud, these invasion effects may be significant and will need to be corrected before using the well data as a basis for seismic modelling.

Lacustrine

A lacustrine system is one where sediment deposition occurs in a lake. At the present day, only about 1% of the Earth's land surface is covered by lakes, though inland seas such as the Black Sea have been large lakes at times of lowered sea-level.

Levee

Levees are ridges built on either side of a river channel. They are typically formed from coalescing crevasse splays, and consist of fine-grained sands and silts. Similar features are seen along the distributary channels of a delta plain, and along deep-water channels of a submarine fan.

Loop

A seismic loop is a single wiggle of a seismic trace, from one zero-crossing to the next.

Migration

Suppose we make a set of seismic records across an area by keeping the source and receiver together and moving the combined source–receiver point around on a regular grid. We could then plot the recorded seismic traces vertically downwards at the proper position on a map of the grid, creating a 3-D volume of seismic traces. This would not give us a correct picture of subsurface reflector geometry, because the reflection points are not in reality vertically below the source–receiver point. If we traced rays from a source position, propagating in all directions into the subsurface, we could find the one that hits a given reflector at right angles. This ray will be reflected back, exactly retracing its path, until it arrives at the receiver. We could also determine the time that the ray would take along this path. Now, what we would like to do is to rearrange the traces in our 3-D volume so that the reflected signal at this travel-time, on the as-recorded trace plotted below the receiver location, is moved laterally and vertically to the real location in space of the reflection point. Migration is this process of moving the as-recorded data to the correct location in space. In reality, of course, seismic data are recorded with a range of source–receiver separations. Ideally, each one should be migrated separately, although it is common to cut down on the computation effort by *stacking* data before migration, which transforms them to the travel-time that they would have for zero separation between source and receiver and then sums them. There are further complications in practice because there may be several rays from one surface point that hit a given reflecting surface at right angles, perhaps in widely separated locations.

Migration aperture

In order to get a satisfactory subsurface image from the migration process, it is necessary to have available a volume of traces surrounding the point at which we want the image. One way of implementing migration is to use the Kirchhoff approach described in section 1.2; for every point in the output image, we want to sum data along a hyperboloid surface in the unmigrated data set. The aperture is the lateral extent of the traces that we take into this summation. Clearly, a very small aperture (a few traces) will do little to reposition data; on the other hand, a large aperture will include very distant traces that will have little influence on the result because the signal at long travel-times will be small. A rule of thumb is that the aperture needs to be twice the lateral distance over which reflections will be moved; a larger aperture is therefore needed for steeper dips, particularly on deep reflectors. When planning a 3-D survey, it is important to acquire the data around the edge of the area of interest that

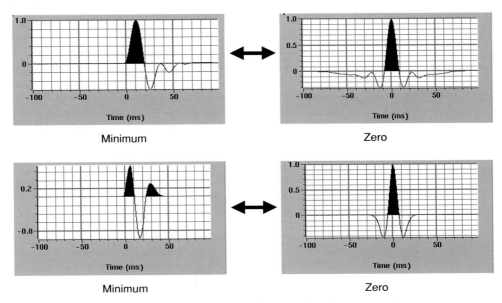

Fig. A2.2 Two examples of zero- and minimum-phase wavelets. In each pair, the zero- and minimum-phase wavelets have the same amplitude spectrum.

will be needed for the migration aperture. This might typically add a 2 km fringe on each edge of the area that is to be well imaged. This makes small 3-D surveys relatively expensive; to image an area 9 km × 9 km (81 km²) would require acquisition over 13 km × 13 km (169 km²).

Migration velocity

A velocity field used to migrate the seismic data to obtain a well-focussed and correctly located image. This is usually closer to seismic velocities in the real earth than stacking velocities, particularly where the velocities have been determined in the course of pre-stack depth migration which allows for the complexities of ray-bending in the overburden. It may still be quite strongly affected by anisotropy, however, and not be well suited to depth-converting picked horizons without further adjustment to tie the well data; lateral resolution of the velocity field may also be an issue for detailed well ties.

Minimum phase

There is an infinite number of seismic wavelets that share the same amplitude spectrum. One of these is the minimum-phase wavelet, which is constructed so as to start at zero time and then have as much energy as possible at the earliest times. In practice, the wavelet will have a maximum value in the first or second loop (fig. A2.2). This type of wavelet is close to what is often generated by real physical sources. Sometimes seismic data are processed to minimum-phase final output. In theory, seismic reflections should then be picked at the zero-crossing corresponding to the start of the reflected signal from the particular interface. In practice, they are usually picked at a maximum excursion on either the first or second loop down from this zero-crossing pick, depending on where the most consistent loop is found. Potentially, this can cause confusion about the polarity of the data if the picking philosophy is not carefully documented. A further disadvantage of interpreting minimum-phase data is that interactions between reflections from closely spaced interfaces are not as easy to visualise as for the zero-phase case.

Moveout

Moveout can refer in several contexts to the way that the arrival time of a signal (e.g. a reflection from a particular horizon) changes systematically across a set of seismic traces. In the particular case where the source–receiver spacing expands symmetrically about a common midpoint (fig. 2.7), then the increase in travel-time from the zero-offset case is called the normal moveout (NMO). Correcting for NMO is essential before traces can be stacked together. This NMO correction will shift the individual samples of a finite-offset trace upwards by an amount that decreases with increasing travel-time. In consequence, far-offset traces will be stretched (NMO stretch) and therefore shifted towards lower frequencies. At shallow depth and long offset, the effect may distort traces so badly that they are unusable and have to be removed (*muted out*) before stack.

ms

Abbreviation for millisecond (1/1000 of a second).

Multiple (reflection)

A seismic signal that has undergone more than one reflection. For example, the signal from a marine source might be reflected from the seabed, reflected again from the sea-surface, then travel down to a deep interface where it would be reflected back to a surface receiver. It will obviously arrive later than a signal that has gone straight from source to deep interface to receiver (the primary reflection). Similar multiples can be created by bouncing the signal between any two interfaces overlying the deep interface, and there could be two or more bounces between these shallower levels, not just one. Deep interfaces can therefore easily become obscured by multiples of interfaces lying above them, if the primary from the deep interface arrives at the same time as the multiple of the shallower interface. For this reason, elimination of multiples is often a key objective of seismic processing.

Mute

Reflection traces recorded at long offset and short travel-time will be strongly contaminated by various types of unwanted signal, such as refractions, and will be distorted by the application of NMO correction which stretches the individual loops. They are usually removed before stack by setting to zero all trace values for offsets beyond a specified offset–TWT curve (the mute). Sometimes an inner trace mute is employed also; this sets to zero all trace values for offsets less than a specified offset–TWT curve, usually to remove traces heavily contaminated by multiples, which are often poorly handled at short offsets by demultiple procedures that rely on NMO differences between primaries and multiples.

NMO

See **Moveout**.

Offset

Source to receiver distance.

Pay (zone)

Hydrocarbon-bearing reservoir; in geophysical discussion, usually irrespective of whether hydrocarbons are producible, economically or at all.

Phase spectrum

When a signal is described as the sum of a series of sine waves by the methods of Fourier analysis, each component sine wave has an amplitude and phase. The amplitude spectrum defines the peak-to-peak amplitude of each sine wave as a function of frequency. However, this is not enough to define the signal. We need also to know what the time alignment of the various sine waves should be. This might for example be seen in the time of the first maximum after time zero. The phase spectrum defines this alignment. A phase of zero means that the component sine wave has a maximum at zero time, a phase of $90°$ means a zero-crossing at zero time, and a phase of $180°$ means a minimum at zero time. The phase spectrum is a plot of the phase of the component sine waves against frequency.

Point bar

Where a river bends, the maximum flow velocity is close to the outer bank. At the inner bank, the flow is less and sediment accumulates to form a point bar, that grows by lateral accretion. Typically the deposits are sands, perhaps with some mud in the upper part.

Primary (reflection)

Signal that has travelled direct from source to reflecting interface to receiver, in contrast to multiples (q.v.) with their more complex paths involving multiple bounces. The primaries carry the information we need to create an image of subsurface structure.

Receiver gather

A collection of the traces recorded at a given receiver from all the various shot points that have been recorded at that receiver. This is sometimes called a *common-receiver gather*. Creation of such a gather involves re-ordering the traces recorded in the field, which will be organised as common-shot gathers, i.e. the collection of traces recorded at all the different receivers from each shot.

Reflection coefficient

When a seismic wave of amplitude A is incident on an interface between two different media, it is in general partly reflected and partly transmitted. If the amplitude of the reflected wave is R, then the reflection coefficient is defined as the ratio R/A. A negative value indicates that the reflected wave is $180°$ out of phase with the incident wave.

Refraction record

To correct land seismic traces for static shifts generated by lateral variations in the near-surface structure beneath shots and receivers, we need to know the thicknesses and velocities of the near-surface layers. Usually, there is a low-velocity weathered layer near the surface, overlying a higher-velocity layer. To investigate the thickness of the low-velocity layer, we can shoot refraction profiles. These consist of long lines of receivers with a source at each end. With this geometry, the first arrival at each receiver will usually be the head wave, often called a refraction. This travels along the top of the high-velocity layer. Geometrically we can think of it as being predicted by Snell's Law, which says that a ray travelling through the interface will be bent (refracted) away from the interface normal. As the angle of incidence of the ray is increased, there will come a point, the critical angle, at which the ray bending would make the ray in the second medium travel along the interface. (The full theory that explains the amplitude of the head wave and its spatial variation is much more complex.) By analysing the variation in head-wave arrival times from one receiver to another, it is possible to map the changes in thickness of the low-velocity layer above the high-velocity refractor.

Root-mean-square (rms) average

The root-mean-square (rms) average of a set of numbers is the square root of the arithmetic average of their squares.

Seismic waves

The most important type of seismic wave is the P-wave, which is an ordinary sound wave. As it moves through the rock, individual particles move backwards and forwards in a direction parallel to that of wave propagation. The other type of wave that can exist in the body of the rock is the shear or S-wave, in which particle motion is perpendicular to the propagation direction. Both types have velocities that depend on the elastic moduli and density of the rock:

$$V_p = [(k + 4\mu/3)/\rho]^{1/2}$$
$$V_s = [\mu/\rho]^{1/2}$$

where k is the bulk modulus, μ is the shear modulus, and ρ is the density. Other types of seismic wave exist, confined to the vicinity of layer boundaries, but they are not important to the interpreter of seismic data.

Shear modulus
Shear deformation of a material involves change in shape without change in volume. The shear modulus of a material is a measure of its resistance to shear stress. For liquids, which will flow freely to accommodate any change in shape of their containing vesssel that does not involve change in volume, the shear modulus is zero.

Stack
In general, the adding of a number of traces together to improve signal to noise ratio. Most often used to refer to the adding of traces with different source–receiver offsets but a common midpoint (fig. 3.7). The traces are first corrected for the increased travel-time at the longer offsets due to the oblique travel path; this is the Normal Moveout (NMO) correction. They may also be corrected for effects of subsurface dip and lateral velocity variation, by some form of migration process.

Stacking velocity
The velocity field that, when used to calculate NMO correction, gives the best alignment of the traces across the CMP gather and therefore the highest amplitude in the stacked trace. It is only loosely connected to the actual velocities of seismic waves in the earth, owing to effects of dip and curvature of the reflector and the impact of lateral variations in the overburden.

Static correction
A static correction is a time shift that is applied uniformly across a particular trace. For example, the effect might be to shift an entire trace downwards by 8 ms. A neighbouring trace might have a different time shift applied. Such shifts are typically needed when processing land data, to remove near-surface time delays particular to each shot and receiver location, as a result of changes in elevation and thickness changes in a near-surface low-velocity layer.

Stratigraphic trap
The simplest sort of oil or gas trap is a domal anticline, or four-way dip closure. Variants on this may require an element of sealing along a fault, but the possible existence of such a *structural trap* can be inferred from the top reservoir map alone. Sometimes, however, a trap requires an element of lateral lithological change to work. This might, for example, be lateral transition from sand to shale within the reservoir formation. Such a *stratigraphic trap* cannot be found from the structural map alone; the lateral change in rock properties needs to be present. It may be inferred from geological argument alone, but seismic evidence for the required transition will reduce the (normally high) risk associated with such a prospect.

Strike section
Seismic section shot perpendicular to the dip direction of the main reflectors, often sub-parallel to the main faults and therefore hard to interpret on 2-D seismic sections because sub-horizontal reflections at nearly the same travel-time may relate to different fault blocks on either side of the line or directly below it.

Trace

Graph showing the amplitude of a seismic signal against time, conventionally plotted with time increasing vertically downwards. The signal can be of many different types, e.g. as-recorded, stacked, or migrated. Originally the amplitude of a trace would be shown by a conventional wiggly line, but increasingly colour is used to convey amplitude information, as explained in section 3.2.

Transgression

Landward migration of the shoreline, owing to relative rise in sea-level.

Turbidites

A turbidity current is a suspension of sediment in a turbulent water flow; such currents are able to move coarse-grained sediment far out to sea and into deep water. Turbidites are the deposits of these turbidity currents. They are widespread deep-water deposits, with individual beds up to several metres in thickness, and ranging from coarse- to fine-grained sediment.

TWT

The Two-Way Time to a seismic reflector is the time taken for a seismic signal to travel from the surface to the reflector and back to the surface again. This is the usual vertical scale for seismic section display.

Unconformity

An unconformity is a surface across which there is a gap in sediment deposition; this may be a result of erosion or of non-deposition. Where the time-gap is substantial, the properties of the sediments are often quite different on either side of it, giving rise to a prominent seismic reflection. Sometimes the unconformity surface cuts across bedding planes of the sediment below it; such angular unconformities are often easily recognised on seismic displays.

Well synthetic

To make it easier to tie seismic to well data, it is useful to make a well synthetic. From the wireline log data in the well, acoustic impedance is calculated as a function of depth by multiplying together the recorded velocity and density logs. From this a reflectivity log is calculated, and an expected seismic response calculated by converting it into a function of TWT using checkshot information if available, and then convolving it with a seismic wavelet. See section 3.1 for details.

Zero-offset

A zero-offset trace is one recorded with source and receiver at the same location. This implies a simple seismic geometry in which an outgoing ray strikes a reflector at right angles and is reflected back along exactly the same path as it has come.

Zero-phase

There are an infinite number of seismic wavelets that share the same amplitude spectrum. One of these is the zero-phase wavelet, which is symmetrical about zero time. In practice, the wavelet will have a strong central loop and a number of smaller sidelobes (fig. A2.2). This is not a wavelet that could be generated by any real source, because it would need to begin before the source was triggered at time zero. However, the seismic traces generated by the real source wavelet can be manipulated by processing into the form they would have if the wavelet were zero-phase. This is useful because it is much easier to understand a zero-phase seismic section. Every reflecting interface produces a signal with its maximum value centred on the interface. This means that a picked horizon representing an interface follows a maximum loop excursion (positive or negative depending on the impedance change), which makes autotracking simple. Where reflections are closely spaced, it is much easier to visualise how they interact to modify the seismic amplitudes when data are zero-phase.

Zoeppritz equations

These equations determine the amplitudes of the reflected and refracted P- and S-waves generated when a plane P-wave is incident on a plane interface between two media of different density and elastic moduli.

Index